A Mixed Bag

Jokes, Riddles,

Puzzles and Memorabilia

Raymond M. Smullyan

Other Books by Raymond Smullyan

A Beginner's Further Guide to Mathematical Logic (2016)
Reflections: The Magic, Music and Mathematics of Raymond Smullyan (2015)
The Magic Garden of George B. And Other Logic Puzzles (2015)
A Beginner's Guide to Mathematical Logic (2014)
The Godelian Puzzle Book: Puzzles, Paradoxes and Proofs (2013)
King Arthur in Search of his Dog (2010)
A Spiritual Journey: Reflections on the Philosophy of Religion, A Transcendental Journey, and Cosmic Consciousness Redux (2009)
Rambles Through My Library (2009)
Logical Labyrinths (2009)
In Their Own Words: Pianists of the Piano Society (with Peter Bispham) (2009)
Who Knows?: A Study of Religious Consciousness (2003)
Some Interesting Memories: A Paradoxical Life (2002)
The Riddle of Scheherazade (1997)
Set Theory and the Continuum Problem (1996)
Diagonalization and Self-Reference (1994)
Recursion Theory for Metamathematics (1993)
Satan, Cantor and Infinity (1992)
Gödel's Incompleteness Theorems (1992)
Forever Undecided (1987)
To Mock a Mockingbird (1985)
5000 B.C. and Other Philosophical Fantasies (1983)
The Lady or the Tiger? (1982)
Alice in Puzzle-Land (1982)
The Chess Mysteries of the Arabian Knights (1981)
This Book Needs No Title (1980)
The Chess Mysteries of Sherlock Holmes (1979)
What Is the Name of This Book? The Riddle of Dracula and Other Logical Puzzles (1978)
The Tao is Silent (1977)
First-Order Logic (1968)
Theory of Formal Systems (1961)

Set in Williams Caslon Text with LaTeX.
Cover Design by Anne Marie Hantho.
Printed in the United States of America.

ISBN: 978-0-9861445-7-8 (paperback)
ISBN: 978-0-9861445-8-5 (ebook)
Library of Congress Control Number: 2016902980

Sagging Meniscus Press
web: http://www.saggingmeniscus.com/
email: info@saggingmeniscus.com

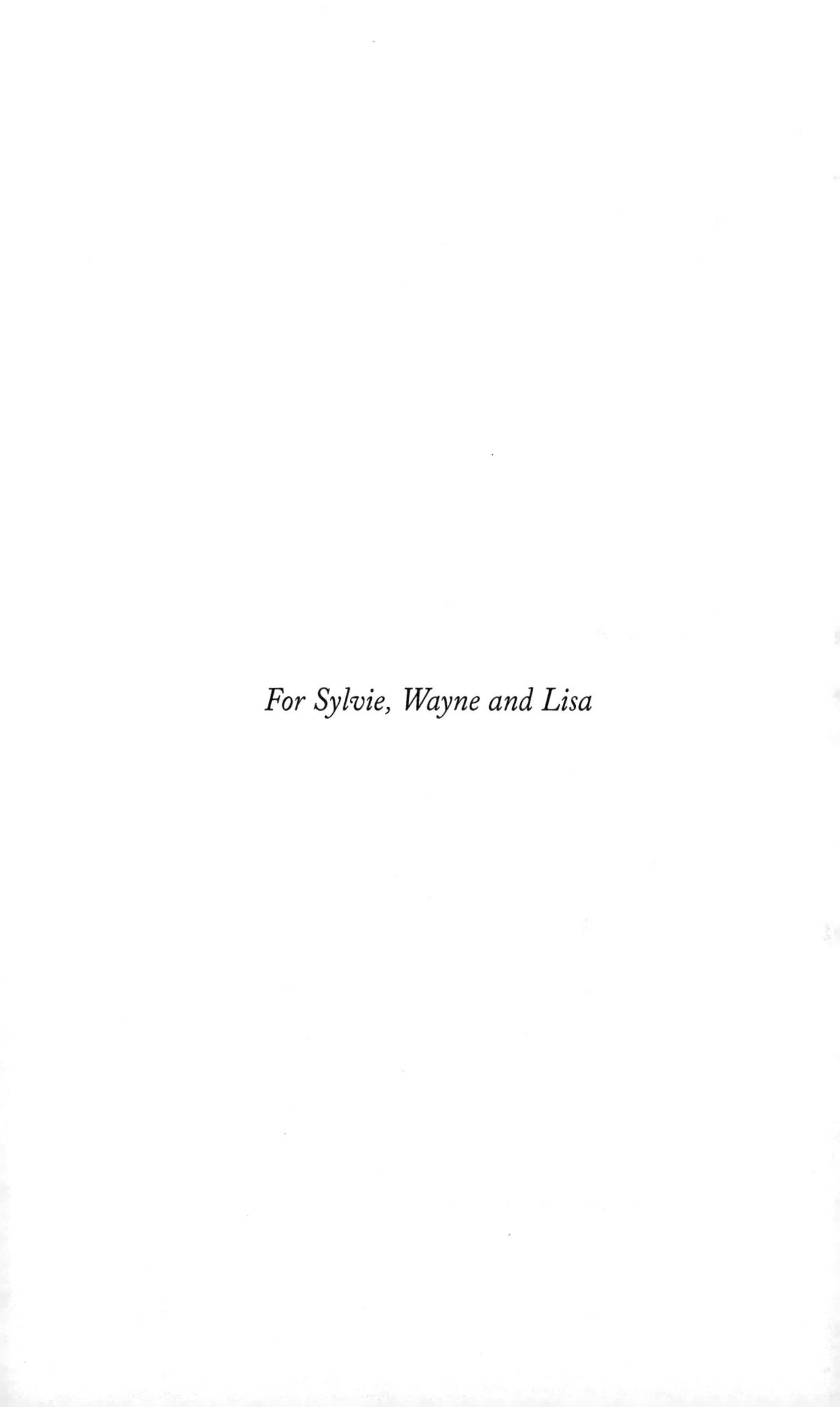

For Sylvie, Wayne and Lisa

Those who cannot appreciate Ray's jokes do not know how to take him seriously.

—Professor Saul Gorn (Computer Scientist)

Table of Contents

Foreword	xi
Preface	xiii
A Mixed Bag	1

Foreword

his book doesn't really need one.

Preface

Doesn't the foreword raise a paradox? It says that it is not needed, but isn't it needed to inform the reader that it is not needed? It if weren't there, how could the reader know that it wasn't needed?

This is reminiscent of an occasion in which I was teaching a graduate course in logic, and before the students came into the room, I wrote the following on the blackboard:

> PLEASE DO NOT ERASE, BECAUSE IF YOU DO, THOSE WHO FOLLOW WON'T KNOW THAT THEY SHOULDN'T ERASE.

A colleague of mine told me that one while he was driving in Canada, he came across a sign that read:

> PLEASE IGNORE THIS SIGN.

I have used a similar gag several times and have sent emails to several friends which read:

> PLEASE IGNORE THIS MESSAGE.

Also, often when I give one of my books to a friend, I inscribe:

> I wish you the best, but I refuse to sign this book! —*Raymond Smullyan*

One of my publishers to whom I gave a book with that inscription cleverly responded by email:

> I appreciate your dedication, but I refuse to acknowledge it.

Many people have told me that my jokes have a special quality not easily found elsewhere, and that I should write a book of jokes. Well, I finally decided to write not a whole book of jokes, but a book in which jokes play a large part, but should also include some of my favorite anecdotes, riddles, puzzles, perhaps a bit of biography, some philosophical discussion, and whatever else comes to mind.

I will not be at all systematic, but just ramble along (as I often do), and will not classify my jokes into groups like restaurant jokes, tailor jokes, Jewish jokes, musical jokes, etc., although my jokes will naturally fall somewhat into groups—for example, if I tell a restaurant joke, other restaurant jokes will come to mind, or if I tell musical jokes, others will come to mind, etc., but this will be spontaneous, not deliberate.

I believe you will enjoy this book more if you do not read too much at a time. Just go and savor it slowly like good tea.

Elka Park, 2009

A Mixed Bag

Some time ago, I got hold of a book of 300 jokes. Of all the three hundred, there was only one I halfway liked; the rest I found totally worthless. Perhaps some of them will appeal to those whose sense of humor is different from mine, I don't know. But for me, there was only one that was not half bad, which is about a conductor on a train who said to the brakeman: "There is a tramp in the box car; throw him off!" The brakeman said, "Certainly," and went into the box car, where he did see a tramp, and said to him, "Now, look, I don't want any argument!" Upon which the tramp pointed a gun at him and said: "This does all my arguing for me!" The brakeman scratched his head, left and went back to the conductor who asked: "Did you throw him off?" "No, I didn't." Angrily the conductor asked: "Why not?" "Well, you see, it turned out that he is my cousin, and I can't throw my own cousin off the train!" "Well, I'll throw him off!" at which the conductor left, and returned after a while: "Well," asked the brakeman, "did you throw him off?" "No," replied the conductor, "it turned out that he's my cousin also!"

Not too bad, but not as good as the average of the jokes I will tell you. Speaking of averages, I am reminded of a joke I heard about sixty years ago: A man took a train from New York to San Francisco. Before getting off, he said to the porter: "What is your average tip?" The porter replied: "Two dollars." He was then given two dollars, and the porter said: "Man, you are the first one to come up to my average!"

And speaking of tips, I am reminded of the gag of Groucho Marx on board ship, when the bellboy had just helped Groucho with his luggage, and Groucho asked him: "Is tipping allowed on this boat?" "Oh, yes Sir!" was the enthusiastic reply. "Well, do you have change of ten dollars?" "Oh, yes Sir!" "Then, in that case, you won't need the nickel I was going to give you."

Speaking of stinginess, I recall one movie featuring Jack Benny and Fred Allen. Jack Benny was portrayed as the ultimate in stinginess. In one scene, Fred visits Jack in one of Jack's mansions, and at one point Jack asks Fred if he would like a cigarette. When Fred says he would, Jack says: "You'll find a cigarette machine in the hall."

There is also the story of a very rich but stingy man who tries to get into Heaven. Saint Peter asks him what he has ever done for anyone. The man replies that he once give a nickel to charity, once gave a nickel to the Salvation Army, and recently gave a nickel to a beggar. St. Peter turns to God and asks: "What should I do with this man?" God replies: "Give him back his fifteen cents and tell him to go to Hell."

One of my favorite stories in this genre is about a very wealthy man who never gave to charities. One day a group of men came to his house and told him he should contribute to United Charities. The man replied: "Just a minute, you haven't heard my side of the story! I have a mother who is very sick and whose medical bills cost a hundred and fifty thousand dollars a year. I have an uncle who is even more sick, and his medical bills cost two hundred thousand dollars a year. My son's college expenses run about eighty-five thousand dollars a year. Now, since I don't give a penny to any of those people, why should I give anything to you?"

There is a certain country in which the inhabitants are characterized by their stinginess. I recall two jokes about this country. One is about an inhabitant who was in a grocery store and after selecting some goods, gave the grocer a dollar bill. The grocer gave him his change. The man held the change in his hand and kept looking at it. The grocer said: "What's the matter, didn't I give you enough change?" The man replied: "Barely."

The other story is about an incident from about eighty years ago. A married couple of inhabitants of this country passed a private airfield and saw a sign saying that the pilot would take one on a ride for half an hour for twenty-five dollars. The husband asked the pilot: "Could you take us for a quarter of an hour for twelve and a half?" The pilot replied that half-hour rides were standard procedure. The man replied: "But if half an hour costs twenty-five dollars, than a quarter of an hour should be twelve and a half." The pilot again told him that the standard procedure was a ride for half an hour, and the two kept arguing. Finally the pilot said: "Look, I can't stand hearing you any longer! I'll tell you what: I'll take you up for half an hour, but I don't want to hear any sound out of you! If you make no sound, I'll charge you only twelve and a half dollars, but if you say just one word you pay me twenty-five!" The man replied, "Fair enough." And so the pilot took them up and the man was quiet for quite a while. The pilot began fearing that he would be losing some money, and so he decided to frighten the man into saying something and made a nose dive and almost hit the ground before flying up again, but no reaction came from the man. Then the pilot made a loop-de-loop and tried one scary trick after another, but still the man was silent. When they finally landed, the pilot said: "Very well, you win—you have to pay me only twelve and a half, but you have amazing self-control! At times it must have been very difficult!" The man replied: "Yes, it was very difficult. It was especially difficult when my wife fell out of the plane!"

The following incident sounds like a Scotch joke, but is actually true! I have a letter of a Scotsman who read one of my books, and wrote me saying: "I very much enjoyed your book; I'm even thinking of buying a copy."

Yes, I really have that letter.

There are, of course, many lawyer jokes such as: "Why is it safe for lawyers to swim in shark-infested waters?" Answer: Professional courtesy.

My favorite one is about a lawyer who said to a client: "You can ask me some questions and I will answer them. I charge one hundred dollars per question." The client said: "Isn't that rather excessive?" The lawyer replied: "I don't think so. Now, what is your second question?"

This reminds me of an allegedly true story of a patient who said to a psychiatrist: "If you help me, doctor, I'll give you every penny I possess!" The doctor replied: "Thirty kronen will be enough." The patient replied: "Isn't that rather excessive?"

I love the scene in a Marx Brother's movie in which Groucho was having dinner with a girl in a restaurant, and after the meal was over, he picked up the bill and said: "This bill is outrageous! If I were you, I wouldn't pay it."

Another Marx Brother's gag is one in which Groucho said: "I could dance with you till the cows come home, or maybe I'll dance with the cows till you come home."

Then, of course, there is the famous one in which Groucho was invited to join a certain club and said: "I would never join any club that would have me as a member."

I equally like Chico Marx. In one scene he tells the story of how he as an aviator tried to fly to France: "I getta half way across, but I have to go back because I no have enough gas. The nexta time I take more gas an get only a mile away from France, but have to come back because I no have enough gas. The nexta time I take plenty of gas, plenty of gas. Halfway across, I have to come back—I forgot the aeroplane."

Another comedy team I like is Laurel and Hardy. In one movie Laurel says to Hardy: "Ollie, I think there is something wrong with your eyes. You should go and see an optimist!"

I am reminded of a father who said to his son: "I think you are seeing double!" The son replied: "Impossible, father. If I were seeing double, I would see four moons up there instead of two."

Coming back to Laurel and Hardy, in one scene, Laurel visits Hardy lying in traction in a hospital and says: "Hello, Ollie, I brought you a present." Hardy replies: "What did you bring me?" Laurel says: "A loaf of bread." Hardy disdainfully says: "A loaf of bread! What kind of a present is that for a sick man?" Laurel replies: "Well, I would have brought you candy, but candy was more expensive."

I like the one about a man who decided to test his wife's hearing, and so when her back was turned, he asked: "Can you hear me?" No answer. He took a step closer and repeated: "Can you hear me?" Still no answer. He came closer still and shouted in her ear: "CAN YOU HEAR ME?" She replied: "I already said *yes* twice before."

I am reminded of the man who went to a doctor complaining that he couldn't remember things too well. "Really?" said the doctor, "how long has this been going on?" The man looked at the doctor blankly and asked: "How long has *what* been going on?"

→ A psychiatrist's secretary came into his office and said: "There's a man in the waiting room who says he is invisible!" "Tell him I can't see him now."

A man came to a psychoanalyst with the following problem: "When I'm in bed, I'm afraid someone is under the bed, and so I go under the bed. Then I'm afraid someone may be on top of the bed, so I go back on top. Then I'm afraid someone is under, and so I go back under, and thus keep changing all night from under the bed to on top of the bed, and hence I can't get any sleep!" The psychoanalyst said: "Your problem is curable. You will have to come five times a week at $150 a session for three years." The man said that he would have to think about it.

Several months later, the psychoanalyst met the man on the street and asked him what he had decided to do. "Oh," replied the man, "I'm completely cured! I told my bartender my problem, and he said that for ten dollars he would tell me how to cure it." "And what was the cure?" asked the analyst. "He told me to saw off the legs."

Speaking of doctors, a woman had a swollen ankle and phoned her doctor for advice. "Put on hot compresses," he said. She put on hot compresses, and the swelling got much worse! Her maid saw what was going on and asked her what she was doing. "I have a swollen ankle," she replied, "and I am putting on hot compresses, as my doctor advised." "No, no," said the maid, "for a swollen ankle, you should put on *cold* compresses, not hot ones!" The lady then put on cold compresses, and the swelling went down immediately. She angrily phoned her doctor and said: "What is this? I put on hot compresses as you advised, and the swelling got much worse. Then I spoke to my maid, who told me to put on *cold* compresses, which I did, and the swelling was enormously relieved!" "That's funny," said the doctor, "*my* maid told me *hot* compresses!"

Speaking again of psychiatrists, one was called into a house in which the problem was that the little girl wouldn't come out from under her bed. Even food was not sufficient to tempt her. Well, the psychiatrist went into the girl's bedroom and came out several minutes later with a puzzled expression. "What have you discovered?" asked the mother, anxiously. "It is very strange," said the psychiatrist, "she is afraid that if she should dare to venture out of her hiding place, people around would start *biting* her!" "Oh, is that all?" asked the mother, "in that case, we'll have to stop biting her."

Then there is the one about a man who came to a psychiatrist and claimed to be a dog. "How long have you believed this?" asked the psychiatrist. "Oh," said the man, "ever since I was a puppy."

Then a man came to a psychiatrist claiming he was dead. "Oh, yes doctor! I'm completely dead!" "Now look," replied the psychiatrist, "I want you to go home and for several times a day, say: *Dead men don't bleed! Dead men don't bleed!* Say this over and over again, and come back to me a month later, and then I'll show you something very interesting indeed!"

The man went home and returned a month later. "Well," said the doctor, "what have you been saying to yourself?" "Oh," replied the man, "*Dead men don't bleed*, doctor; *dead men don't bleed*." "Good," said the doctor, "now watch this!" At which he pricked the man's finger with a needle, and the man saw it bleed. "Oh, my God," said the man, "my God! Dead men *do* bleed!"

The wildest psychiatrist story I know is about the man who came to a psychiatrist with the complaint that his whole family was plotting against him. The psychiatrist replied: "Perhaps you should kill them!"

The following story is true: a psychiatrist I know worked a great deal with psychotics. At one point he came into a room in the hospital and saw a man with his ear to the wall. "What are you doing?" the doctor asked. "Ssh, doctor, I'm listening!" "And what do you hear?" The patient replied in a pathetic voice: "That's just it, doctor, *nothing*!"

Another true incident is about a schizophrenic patient who hadn't talked for sixteen years. He was being exhibited to a medical class. One of the students went up to him and said: "Why don't you ever say anything?" The patient opened his mouth and said: "What is there to say?"

Another true story beautifully illustrates a point that I have emphasized in some of my other writings, namely that a lie doesn't necessarily involve making a *false* statement, but only that the statement be contrary to the person's belief, even if the statement is true. Well, the doctors at a certain mental institution were thinking of releasing a certain schizophrenic patient, but they decided to first test him under a lie detector. At one point they asked him whether he was Napoleon. The patient replied: "No." The machine showed he was lying!

Speaking of lying, the following incident is true. During part of my college years, I was supporting myself as a magician. At one point my business was not going too well, and so to earn extra money, I applied for a job as a salesman. I had to take an examination, and one of the questions asked was whether I objected to telling a little lie now and again. I certainly did object, but I was afraid that if I truthfully voiced my objection, I wouldn't get the job, and so I lied and said "No."

Later on, I realized I was in a kind of paradox! Did I object to the lie I told the sales company? I realized that I did not! Then since I didn't object to that particular lie, it therefore followed that I *don't* object to all lies, hence my answer "No" was not a lie, but the truth! So was I lying or not?

To my utter amazement I once came across a psychiatry textbook written around 1902, in which schizophrenia was defined as: "A mental disease caused by disobeying one's parents." There may actually be a germ of truth in that!

Someone I know told me that he once visited a mental institution and spoke to a patient who complained that his whole family was plotting against him. When asked how he knew, he replied: "Because when I went to the gas station, I could see though the keyhole..." "Just a minute," said my friend, "how could you see through the keyhole if you went to the gas station?" "Ah," he replied, "that's what I told them!"

There is the story told about the wife of the governor who visited a mental institution and met a patient who said: "Now look, I can assure you that I have been framed by my family. I understand that you don't necessarily believe me, since other patients here make the same claim. No, I am not advising you to believe that what I am saying is necessarily true, but I know that you are the wife of the governor, and all I am asking is that you ask the governor to have my case received. That's all I am asking." This sounded reasonable to the lady and she agreed to do this. She turned her back and started walking out of the room, at which the patient gave her a terrific kick in the backside and yelled: "And don't forget to tell the governor!"

Two patients of a mental institution were standing outside in the garden with a warden. A bird flew by and dropped something on the head of one of them. The warden, who was quite kind-hearted, said: "Oh, I'll go inside and get some toilet paper." When he went inside, the patient laughed and said to the other one: "That guy is crazy!

By the time he gets back, the bird will be twenty miles away!"

A man once had a flat tire just as his car was outside an asylum. He removed the wheel, but by accident, all the lugs rolled away and down a gutter! The man scratched his head and wondered what to do. At this point a patient, who had been watching through an open window on the second floor, said "why don't you take off one lug from each of the other three wheels and use them for this one. Then, at the next gas station, you can probably get four more." "What a clever idea!" said the man, "How come you are here?" "Look," said the patient, "I may be crazy, but I'm not stupid!"

There is the allegedly true story about a truck that was a bit too high for a tunnel and got stuck in it and couldn't go either backwards or forwards. A crowd gathered discussing what should be done. One little girl of four years old kept saying: "I know what to do!" They all told her to keep quiet. Finally, one of the crowd asked her: "Okay, what would *you* do?" She said: "I'd let some of the air out of the tires."

Now for a little puzzle: A little girl said to her mother: "Let's you and me go to the railroad station and meet Daddy and the four of us will go home to dinner." Why did she say *four* instead of *three*? (The answer is given after the next item.)

Another puzzle: A certain man had great grandchildren, yet none of his grandchildren had any children! How is this possible?

Answers:

(1) She was too young to count correctly.
(2) He had grandchildren who were great!

Another true story: One of my wife's grandchildren at the age of five once said: "I hope I never get to be ninety-nine!" When asked why, he replied: "Because when you're that old, you could die!"

At the age of three, another of my wife's grandchildren was about to be taken up in a plane by his father, and asked: "Daddy, when we go up, will we also get small?"

Some riddles:

a. Where do departed spirits get their mail?
b. Where are the letters kept?
c. Where is it located?
d. What is the zip-code?
e. When a book is tired, why should it be sent to Romania?

Answers:

a. At the ghost office
b. In the dead letter department
c. In Death Valley
d. Either 00000, or infinity; I'm not sure which!
e. To give the book-a-rest.

A Buddhist came to Hungary in the Middle Ages and tried to convert everyone to Buddhism. He was really quite aggressive and made a real pest of himself! He was subsequently knows as the Buddha Pest.

In the Basque region, between France and Spain, there was a small theater with only one exit. On one occasion, the theater was full of Basques, when a fire broke out. They all rushed to the one exit and some of them got trampled! This proves that one shouldn't put all one's Basques in one exit.

A man once came across a girl reading a book on love-making. She turned to him and said: "It says here that the best lovers are American Indians and Poles. Did you know that? My name's Mary, what's yours?" He replied: "Red River Kowalski."

In a small town, one man said to his friend: "Did you know there's a funeral today?" "Yeah? Who died?" "I don't know, I think it's the one in the coffin."

In England, a man met a friend and said: "I hear your wife died." "Eh?" "I said that I understand your wife died." "Eh?" The man shouted louder, "I hear you buried your wife!" "Oh, yes, had to, dead, you know."

When I was in England, I was waiting in London for a bus to Dover. It was raining and I tried to put on a raincoat, but I got my left arm through the wrong opening and got all tangled up. A British lady, seeing my plight, got up from her seat, helped me and said: "Got it wrong, have you?" (How typically British!)

There is the story of the Englishman visiting New York and dining in the Waldorf Astoria. To show how democratic he was, he got into a chummy conversation with the waiter, who liked him, and said, "I have a riddle for you: my parents gave birth to a child and that child was neither my brother nor sister. Who was it?" "Neither your brother nor sister? Why, who could it possibly have been?" "Why, it was *me*!" "Oh, yes! Jolly good joke! Jolly good joke!"

Several months later, back in London, the man was at a party and said: "When I was in the States, I heard a joke. Let's see now, how did it go? Oh, yes; my parents gave birth to a child and that child was neither my brother nor sister. Who was it?" "Who was it?" they all cried; "we have no idea!" The man replied: "Why, it was the waiter at the Waldorf Astoria!"

Another Englishman, when he came to America, visited a food factory and asked one of the workers: “What do you do with all this food?” The worker, who was a bit of a wise guy, replied: “We eat what we can, and what we can’t, we can. Ha! Ha!”

Upon returning to England, he told some friends: “You know, in America, they have a very clever system for producing food. They eat what they can, and what they can’t, they tin.”

I am very fond of British humor. As an example, a commentator of the BBC (British Broadcasting Corporation) was in the process of introducing a modern composer. In a perfectly deadpan manner he said: “He started out life as a dentist. Then he decided to see if he could extract a living from music.”

Another example of British humor is an incident told to an American press conference by the chief inspector of Scotland Yard. To begin with, he spoke about gun control and how well it was working in England. At the end of the hour, after hearing all about how well it was working, some idiot American reporter asked: “Tell me, Inspector, do you believe in gun control?”

Anyway, the inspector then told the delightful incident of a criminal investigator who was testifying before a judge about a matter pertaining to chemistry. When he finished, the judge said: “After having heard what you said, I am no wiser than before!” The investigator replied: “That may be true, your honor, but you are better informed.”

When I was in England, I proved to a famous British logician that in England, they drive on the wrong side of the road. My proof was this: In America, we drive on the right side of the road, don’t we? Well you people don’t drive on the right side, therefore you drive on the wrong side!

The above is a perfect example of equivocation—the fallacy of using the double significance of a word. Another example is in one of Plato's dialogues in which the sophist Dionysodorus proves to one Ctessipus that Ctessipus' father is a dog. The argument is as follows:

> DIONYSODORUS: You say you have a dog?
> CTESSIPUS: Yes, a villain of one.
> D.: And has he puppies?
> C.: Yes, I certainly saw him and the mother of the puppies come together.
> D.: And is he not yours?
> C.: To be sure he is.
> D.: Then he is a father and he is yours, ergo he is your father and the puppies are your brothers.

In another Plato dialogue, Socrates criticizes the sophist Protagoras for taking money from his students for teaching them wisdom. Protagoras then explains that at the end of the course of instruction, if the student feels that he has not learned enough, his money is refunded in full.

Upon reading this, the following funny thought occurred to me: after the course of instruction, a student comes to Protagoras demanding his money back. Protagoras then asks him whether he can give a good argument why he should get his money back. The student then gives an excellent argument, upon which Protagoras says: "See the excellent dialectical skill that I have taught you!"

I told this to my stepson, Dr. Jack Kotik, who suggested the following addition: another student comes to Protagoras demanding his money back, and again, Protagoras asks him whether he can give a good argument to justify his demand. After a pause, the student says "No." Upon which Protagoras says: "O.K., here's your money back."

There is a very nice legal paradox known as the "Protagoras paradox" about a poor but talented student who comes to Protagoras, but has no money to pay for tuition. Protagoras, seeing that the student is talented, makes the following agreement with him: the student pays nothing during the teaching period, but agrees that after the training period, he will pay a certain amount after winning his first case. Well, after the course of instruction, the student doesn't take any cases, hence Protagoras sues him. He gives the following argument: "Either he wins the case or he loses the case. If he loses, then by definition, he must pay me—this is what the case is about. On the other hand, if he wins the case, then he will have won his first case, and so by agreement, he must pay me." The student then said: "He has it all wrong! If I win the case, it means I don't have to pay; that's what winning the case means. On the other hand, if I lose this case, then I will not have yet won my first case, hence I won't yet owe him anything!"

How should the case be decided? The cleverest solution I have yet heard came from a lawyer, who said: "The court should have the student win the case—he doesn't have to pay. Then Protagoras should sue the student a *second* time, since the student would then have won his first case." Clever!

I love paradoxes! One of my favorites is:

YOU HAVE NO REASON TO BELIEVE THIS SENTENCE.

Do you have any reason to believe that sentence or don't you? Most people say they don't. Well, if you have no reason to believe that sentence, then what the sentence says is true, which is good reason to believe it! On the other hand, if you have good reason to believe the sentence, then the sentence is true, which means that you have no reason to believe it! So either way, you're stuck!

Well, I was once brilliantly outwitted about this by a child of 9½ years old! I gave a lecture at a rather prestigious university, and to give the audience something to mull over, I came into the lecture hall a half-hour early and wrote the above sentence on the board in large letters. Then, a half hour later, I came down the stairs to a full audience and noticed this very bright looking kid sitting in the front row, and I asked him: "Do you believe that sentence?" To my surprise, he replied: "Yes." Taken aback, I asked: "What is your reason?" He replied: "I don't have any!" I then asked: "Then why do you believe it?" He replied: "Intuition." He escaped the paradox perfectly!

I once thought of a combination of a paradox and an insult. Consider the following sentence:

ONLY AN IDIOT WOULD
BELIEVE THIS SENTENCE!

Now, imagine two people A and B looking at that sentence written on a blackboard. A asks B: "Do you believe that sentence?" B replies: "Of course not! Only an idiot would believe that sentence!"

Now B is not being logically inconsistent, but he is being what I would call *psychologically* peculiar! By saying "only an idiot would believe that sentence," he is agreeing with the sentence, hence he clearly believes it. But he said that he didn't believe it. And so he is in the strange position of believing something and also believing that he doesn't believe it! Now, this is not a *logical* inconsistency, but it certainly qualifies for what I would call a psychological peculiarity!

Several years ago a colleague of mine was visiting me with his six-year-old daughter Miriam. During dinner, the father did not like the way she was eating and said: "That's no way to eat, Miriam!" She replied: "I'm not eating Miriam!" Pretty clever for a six-year-old, eh?

On another occasion, a friend of mine named Stanley was visiting us with his 14-year-old boy John. As usual, Stanley was talking for about two hours when John said something. At which Stanley said, "Who's talking?" John replied, "You are, Dad; you have been talking for the last two hours." "And whose fault is that?" said Stanley. "Obviously mine," replied John, "since I didn't interrupt you." At this, Stanley (as well as myself, of course) was amused.

I am reminded of a true, but funny story about the philosopher Moses Mendelssohn, the grandfather of the composer Felix Mendelssohn. He was on fairly good terms with Frederick the Great. One morning, on a walk, he met Frederick the Great who said: "Good morning, Master Mendelssohn, where are you going?" Mendelssohn replied: "I don't know." Frederick angrily said: "How dare you give me such an answer? Guard, take him to prison!" At which Moses said: "You see, your Majesty, did I know I was going to prison?" This amused Frederick, who then pardoned him.

Speaking of philosophers, a philosopher once had the following dream: First Aristotle came by, and the philosopher said to him: "Can you give me a fifteen-minute thumbnail sketch of your entire philosophy?" To his amazement, Aristotle did an excellent expository job; he packed in the essential ideas. Then the philosopher had an objection to Aristotle's system that Aristotle could not answer, and so, embarrassed, he melted away. Then Plato came by, and they went through the same routine, and the philosopher had the same objection to Plato's system as to Aristotle's, and again Plato couldn't answer it, and melted away. And then all the philosophers of history, the ancients, the medievals and the moderns, came by, one by one, and the philosopher had the very same objection to every one of their systems, and none of them could answer it. Well, after the last philosopher had disappeared, the dreaming philosopher said to himself: "I know I'm asleep and dreaming all this, but here I have discovered a universal refutation for *all* philosophical systems, and when I wake up tomorrow, I will have forgotten it and the world will miss something extremely important! If only I could wake myself up and write it down!" Well, with an iron effort, he woke himself up, rushed over to his desk, wrote down his universal refutation and went back to bed with a sigh of relief! Then next morning, he woke up and remembered his dream, and so he rushed over to his desk to see his universal refutation, which was: *That's what you say!*

Unlike the above story, the following incident is true: The famous turn of the century American psychologist and philosopher William James, when once asleep, dreamed that he had discovered the fundamental secret of the universe. He woke himself up and wrote it down and went back to sleep. The next morning he read what he had written, which was "Hogamus, higamous, man is polygamous. Higamous, hogamous, woman is monogamous." This is really true, and I think quite revealing!

James once asked a boy: "Do you know what faith is?" The boy answered "Yeah. Faith means believing something you know ain't true."

The philosopher René Descartes began his philosophy with: "I think, therefore I am." I once thought of this variant: "I think therefore I am? Could be, or is it really someone else who only thinks he's me?"

This is reminiscent of the joke about a man who went into a bar and had the following conversation with the bartender:

> MAN: You don't know who I am, do you?
> BARTENDER: No.
> MAN: You've never seen me before.
> BARTENDER: No.
> MAN: But you just saw me walk in.
> BARTENDER: Yes.
> MAN: Then how do you know it's me?

This further reminds me of the man who was diagnosed as schizophrenic. The reason? It is that he kept forging his own signature!

There is the story that Descartes was at a bar. The bartender asked, "Monsieur Descartes, would you like a cocktail?" He replied: "I think not," and disappeared.

A boy once asked his father: "Daddy, what is philosophy?" The father replied: "Philosophy, my boy, is: I think therefore I am." Puzzled, the boy said, "But where does this leave me?"

There is also the story about a boy who asked his father: "Daddy, what is ethics?" The father replied: "Ethics, my boy, is this: The other day a lady came into my store and gave me a $20 bill, thinking it was a ten. I also thought it was a ten, and accordingly gave her change of ten. Later I discovered that it was really a twenty. Now, ethics, my boy, is this: should I tell my partner?"

On a more serious note, speaking of ethics, I have elsewhere said that I am quite concerned about the prevalence of retributive ethics—the belief that evildoers deserve to suffer for their evil. As I see it, all of us—the best of us—have sadistic tendencies—perhaps on a subconscious level—and the only socially acceptable outlet for our sadism is retribution. Although it is considered wrong to inflict suffering on an innocent person, it is right to do so to a guilty person—in fact justice demands it! Well, I don't believe that this is the ethics of the future. I predict that the decline in retributive ethics, the decline of wars and the decline of crime will come to the human race hand in hand.

Enough philosophizing. The last joke I told you is related to the one about the owner of a bar who watched his bartender who, when receiving money from a customer, put some of it into the cash register and the rest into his pocket. One one occasion, he put all the money into his pocket, at which the owner said: "Jake, since when are we no longer partners?"

This is related to the following true incident: A lady I know was a frequent customer of a famous delicatessen in New York City. On one occasion she saw an elderly woman sneak some food into her pocket and walk out without paying for it. The lady went to the owner and told him about this. He replied, "Oh yes! She's actually a good customer, but I know she occasionally steals a little of my supplies." "Haven't you ever spoken to her about this?" my friend asked. "Oh no," he replied, "that would embarrass her terribly!"

The above incident happened over forty years ago. The same lady, who was a baker, went into a grocery story where butter was 73 cents a pound. She took 10 pounds and brought them to the cash register, and explained: "I have here 10 pounds at 73 cents a pound, and so I owe you $7.30." The clerk said: "Just a minute, lady," and wrote down 73 ten times, added them up, and said: "You're right!"

The fact is that five out of four people have trouble with mathematics.

As some of you know, a person who is able to add, but unable to multiply, can multiply if using a log table (table of logarithms). Well, the story is told that when Noah's ark landed, the snake said to Noah: "If you want me to have children, you will have to cut the tree over there into pieces." Noah was quite puzzled, but cut the tree into pieces. Some months later when he came back, sure enough, there were a lot of little snakes. Noah asked the snake: "Why did I have to cut the tree into pieces in order for you to have children?" The snake replied: "Because I'm an adder and I need logs in order to multiply."

There are three kinds of mathematicians in the world—those who can count and those who cannot.

As to philosophers, they can be classified into two categories—those who classify all philosophers into two categories and those who don't.

The philosopher Henri Bergson, who wrote a whole book on the philosophy of laughter, evidently had little sense of humor himself as the following incident will reveal. Bergson was at a party at which was also Bernard Shaw, who was expounding Bergson's philosophy to the group. At one point Bergson interrupted him and said: "No, no, Monsieur! That is not at *all* what I meant!" Shaw replied: "Please, I understand your philosophy much better than you do!" Bergson did not at all see the humor of the remark, and was quite furious!

Do you know the difference between a philosopher and a theologian? A philosopher is one who looks in a dark room for a black cat which isn't there. A theologian is one who looks in a dark room for a black cat which isn't there and finds it!

A certain philosopher once remained in a closet for 25 years, to contemplate what life really *was*! After the 25 years, he went outside and met an old colleague, and the following conversation took place.

> COLLEAGUE: Good heavens, where have you been all these years?
> PHILOSOPHER: In a closet.
> COLLEAGUE: What? Why in a closet?
> PHILOSOPHER: I wanted to discover what life really *is*!
> COLLEAGUE: And have you found an answer?
> PHILOSOPHER: Yes.
> COLLEAGUE: *(eagerly)* And what is the answer?
> PHILOSOPHER: *(gravely)* Well, it can best be described by saying that life is like a *bridge*.
> COLLEAGUE: That's very interesting, but could you be a bit more specific? Could you please tell me just *how* life is like a bridge?
> PHILOSOPHER: *(after a pause)* Er, er, maybe you're right, maybe life is not like a bridge.

There are two variants of the above joke. One is about a man who asked his priest what life really was. The priest sent him to the archbishop to answer the question. Then the archbishop sent him to the Pope. The Pope said: "The only person in the world who knows the answer to *that* is a certain Lama in Tibet." Well, it took the man twelve years to find the Lama, and when he found him, he said:

"I've come to inquire of you, Holy Lama, what life really *is*. What is it?" "Well, my boy," replied the Lama, "I would say that life is like a fountain." "A *fountain*?" cried the man in amazement. "Yes," said the Lama. "Isn't it?"

Then there is the Jewish version: A man comes to his rabbi to ask what life really is. The rabbi replies: "Life is a barrel of herrings." "Great!" said the man, "Wonderful. But could you please tell me just *how* this is so?" "Alright," said the rabbi, "so it's *not* a barrel of herrings."

I am particularly fascinated by similarities of jokes between completely different ethnic groups. For example, here is a Zen story that is quite similar to a Hasidic Jewish joke.

The Zen version is this: A monk came up the mountain to interview the Master, who asked him whether he came from the North or the South. "The South," was the reply. "In that case," said the Master, "have a cup of tea." The next morning, another monk came up the mountain for an interview, and the Master likewise asked him whether he came from the North or the South. This time, the monk said he had come from the North. "In that case," said the Master, "have a cup of tea."

Later on, the Master's assistant said to him: "I don't understand, Master; you told one that since he was from the South, he should have a cup of tea, and the other, that since he was from the North, he should have a cup of tea. How come?" The Master replied: "Have a cup of tea."

The Jewish version is about two women arguing about which one owned a certain chicken. They could not come to an agreement, and so they went to a rabbi to decide the case. His first interview was with the first woman alone and after hearing her case, he said, "I agree with you; you're absolutely right!" Then he interviewed the second woman alone, and after hearing *her* case, he said: "I agree with you;

you're absolutely right!" Later on, his wife, who in the next room had heard all this, said to him: "Now look, how could they possibly *both* be right? If the first lady was right, then the second was wrong, and if the second was right, the first was wrong. They can't *both* be right!" The rabbi replied: "I agree with you dear, you're absolutely right!"

Being Jewish myself, I can tell Jewish jokes without being politically incorrect (although I don't mind being politically incorrect; I think this whole idea of "political incorrectness" is incorrect!). Anyhow, do you know the *correct* definition of a Jew? This definition is one that applies to me perfectly! A Jew is one who upon hearing a joke says: "No, no; it should be told *this* way!" I'm afraid I do that often!

Some people are very bad at telling jokes. A perfect illustration will soon be given. But first there is the following riddle: Upon the 30th floor of an apartment building there lived a man, who every morning took the elevator down to the ground floor, went to work, and came back in the early evening, entered the elevator and pressed the button for the 29th floor, got out and walked up one flight of stairs. The question is: Why did he press the 29th floor button instead of the 30th?

The answer is that he was a midget and couldn't reach the button for the 30th floor. Well, at a party once, someone told this joke in the following manner: "Upon the 30th floor of an apartment building there lived a midget...."

The following incident, which I just recalled, is true: a woman who had just come over from Europe was at a party one evening, and upon leaving, shook the hand of the hostess and said: "I want to thank you for your hostility!"

Speaking of errors, the following cute ones of children have been recorded:

1. With his own hands, Abraham Lincoln built the cabin in which he was born.
2. Bach was the greatest composer who ever lived, and so was Handel.
3. Queen Victoria sat on a throne for 50 years.

Again speaking of errors, can you solve the following puzzle? (Answer is given after the item following this one.)

> There are three errers in this sentense. Can you find them all?

My dear late friend, Professor Saul Gorn, a computer scientist, wrote a delightful, privately published series of sentences that somehow manage to defeat themselves. He titled the whole collection: *Saul Gorn's Compendium of Rarely Used Clichés*. Here are some of my favorites:

1. Before I begin speaking, there is something I would like to say.
2. These days, every Tom, Dick and Harry is named John.
3. Half the lies they tell about me are true.
4. This book fills a long-needed gap.
5. I'm not leaving this party till I get home!
6. If Beethoven were alive today, he would turn over in his grave!

Answer to the "error" puzzle: The first error is that *errer* is obviously misspelled. The second is that *sentense* should have been *sentence*. The third that there there were only two errors, not three!

But doesn't this raise a paradox? Since the third error has just been found, then there really are three errors, hence the word "three" in the sentence is right after all, but then where is the third error? How does one get out of this?

I like the Businessman's Paradox due to Lisa Collier: the president of a certain company offered a hundred-dollar reward to any employee who would make a suggestion that would save the company money. One employee suggested: "Eliminate the reward!"

I also like Quine's paradox of the machine that works only when it is out of operation.

Now let me say a little about computers: do you know the story of the military computer? Well, when the army first sent a rocket to the moon, the Colonel programmed in two questions:

1. Will the rocket reach the moon?
2. Will it return to earth?

The machine thought for a while, and finally a card came out which said *Yes*. the Colonel was furious, since he did not know whether *Yes* was the answer to the first question, the second question, or the conjunction of the two, so he angrily programmed back: "Yes what?" After a while a card came out which said *Yes Sir!*

Then there was the computer that knew *everything*! The salesman was demonstrating it to a customer, and asked him to ask anything of the machine. "Okay," he said, "where is my father?" The machine thought for a while, and finally a card came out saying: "Your father is now fishing in Canada." The customer laughed and said: "You see, the machine is no good! It is happens that my father has been dead for many years!" The salesman said: "No, no; you must ask the machine in more precise language. Here, I'll ask the machine for you." He then asked the machine: "This man here, where is his mother's husband?" After a while, a card came out of the machine saying "His mother's husband has been dead for many years. His father is now fishing in Canada."

There was a computer that was an expert on making money on the stock market! For a fee of $100, you could get advice from the machine. Well, a man paid $100, and then asked the machine: "How do I make money on the stock market?" The machine advised: *Buy low, sell high*.

Someone once asked the mathematician and computer expert Alan Turing whether he believed that computers could think. He proposed the following test for whether or not a machine can think. Suppose you are in one room and in another room is either a human or a computer; you are not told which. Messages can be typed back and forth between the two rooms. Well, if after several hours of this, you couldn't tell which, and if in fact it is a machine in the other room, then, so Turing would say, the machine can think.

This is known as the Turing Test. Personally, I don't believe in the validity of this test for one minute! To me, there is all the world of difference between *simulating* thought and actually thinking, but that is neither here nor there. At any rate, I understand that a computer has been invented that is so remarkably intelligent that if you put it into communication with either a computer or a human, it can't tell the difference!

Another miracle: a certain scientist, whose name I won't mention, has actually invented a means of talking to a man two hours after he is dead! Yes, this is really true! How is it possible? (Solution is given following the next item.)

Another puzzle: The great Harry Houdini did the following spectacular trick: he would enter an oven dressed in only a loincloth with a piece of raw steak. The oven had a glass door through which the spectators could see. Houdini and the steak would remain in the oven for about ten minutes and then upon coming out, the steak was burned to a crisp, but Houdini was unharmed. How did he do it?

Solutions:

1. Anybody can talk to a man two hours after his is dead. The man doesn't talk back, of course!
2. Anybody can do this trick! The fact is that that any *live* human being can stand for a few minutes a temperature sufficient to cook a steak, because alive humans *perspire*! I emphasized *live*, because if a dead human had been put in that oven, it would have been cooked just like the steak. But the perspiration of a live human being is sufficient to prevent severe damage—at least for a time. Personally, I would not like to try this, but Houdini had fantastic endurance!

On the subject of Houdini, I know several choice incidents: Houdini could usually get out of a locked prison cell in about eight minutes. Well he was once outwitted by a British police officer. In London, he was put in a prison cell, and it took him a couple of hours to get out! Reason? The officer *didn't really lock the door*! The door was open the whole time, and none of Houdini's usual tricks would have any effect on an open door!

As many of you know, Conan Doyle in his later years got involved with spiritualism and the supernatural in a completely crazy manner! He *insisted* that the way Houdini got out of locked trunks was by dematerializing and going through the keyhole! Nothing Houdini could tell him could change his mind! He insisted that Houdini was lying, and wrote him saying: "You really should share your secret of dematerialization with the world! It's too valuable to keep for just yourself."

Conan Doyle once attended a mind reading act in London by a husband and wife team. After the performance, Doyle went backstage to congratulate the couple on their psychic powers. The husband replied: "I'm sorry to disappoint you, Sir Doyle, but we don't have psychic powers. We use signals." Doyle angrily replied; "I'm sure you do have psychic powers, whether you realize it or not!" and left the room in a huff.

On another occasion, Conan Doyle and his wife took Houdini to a spiritualistic medium to contact Houdini's departed mother. The medium went into a trance (supposedly) and all these impressive words came out—impressive to the Doyle's, that is. But Houdini was laughing through it all. After the performance was over and the three left, the Doyles asked Houdini why he was laughing. Houdini replied: "If that had been my mother, she never knew English; the only language she knew was Yiddish!"

The following incident is one that I found quite touching! Although Houdini had debunked countless mediums, and had little use for spiritualism, he nevertheless told his wife Beatrice: "I do wish to be open-minded about all this, and so if I should die before you, and there really is an afterlife and any possibility of communicating with the living, I promise you I will do my best to contact you."

Well, he did die before her, and about six months later I saw at the back of a British magic magazine a message to Houdini from his wife which was:

> *Dear Harry: You were right as usual. You didn't come back.*

I was a magician for many years, doing table magic at night clubs and restaurants. On one occasion I came to a table in which there was only one man, smoking a pipe. He was about the most blasé character I ever met; nothing I did impressed him in the least. He just kept smoking his pipe and maintained a completely deadpan expression. I made my tricks better and better, to no avail. Finally, about twenty minutes later I did my most spectacular trick, at which he took out his pipe, slammed the table and angrily yelled, "it's a *trick*!"

People in those days, knowing I was a magician, often asked me if I ever sawed a lady in half. I told them that in my lifetime I have sawn over a hundred women in half, and I am learning the second half of the trick now.

Now let's come to logic. A police officer whom I knew, knowing that I was a logician, told me *his* idea of logic: "My wife and I once came late to a party. There were only two pieces of chocolate cake left on a plate, which the hostess offered us. Since I was closer to her, she offered it to me first. Well, I reasoned as follows: I like chocolate cake, my wife likes chocolate cake, she knows I like chocolate cake, she loves me and wants me to be happy, so I took the larger piece."

This reminds me of the story of two men in a restaurant who ordered fish. The waiter brought two fish on a platter, one large and one small. One of the men held the platter in front of the other and said: "Please, help yourself!" The other said, "O.K." and took the larger piece. There was an angry silence for about a minute, until the one with the smaller piece said: "You know, if you had offered it to me,

I would have taken the smaller piece!" The other one said: "What are you complaining about, you have it, don't you?"

A somewhat similar story concerns a banquet at which asparagus was passed around the table on a platter. When it came to one lady, she cut off all the tips, put them on her plate and passed the rest on to her neighbor, who furiously asked: "Why did you keep all the tips for yourself and pass the rest onto me?" The lady replied: "Oh, the tips are the best part, didn't you know?"

The following incident is true: Two mathematicians were dining in a restaurant. When the meal was over, the waiter came and said to one of them: "Do you want your checks separate or together?" He replied: "Separate." The waiter then said to the other: "You want yours separate too?"

I love illogicalities! Another occurred on a bus about twenty years ago when smoking was allowed only in the last two rows. Well, shortly after the bus started, the driver's voice came over the loudspeaker: "Federal regulations strictly forbid smoking, only in the last two rows."

Once in a restaurant I saw a sign which read: *We cater to schools, clubs and other occasions.*

In another restaurant there was a sign: *Please wait for the hostess to be seated.*

In another restaurant, there was a sign on the outside: *Specials: Mondays, roast beef. Tuesdays, closed.*

In a small town I once visited, there was a big storm. The radio announcer who wished to emphasize its seriousness, said: "This storm has already been attributed to three deaths!"

Another illogicality: A soapbox orator was once decrying the fate of the poor, and said: "And a painter without paint can't paint unless he has canvasses!"

Another: A reporter once gave the following account of an automobile accident involving a woman: "She was physically unhurt, but she was in such a state of shock that she was unable to confuse fantasy with reality."

The following is not an illogicality, but it raises an interesting question: In a restaurant I once saw the following sign: *Good food is not cheap. Cheap food is not good.*

The question is this: Do the two sentences say the same thing or different things? (The answer is given following the next item.)

I was once in a Chinese restaurant with another mathematician. On the top of the menu was printed: *Extra charge for anything served extra*. My friend observed: "They could just as well have left out the first and last words."

Answer to the Question of the Two Sentences: Many people believe that the two sentences say different things, but logically speaking, they say *exactly* the same thing—namely, that no food is both good and cheap. (In general the sentence "No A is B" and "No B is A" are synonymous; they both say that nothing is both A and B.) Although *logically* the two sentences are equivalent, it is understandable that many people think they are different, because *psychologically* they tend to convey quite different images: when one sees "Good food is not cheap" one tends to think of good expensive food, whereas when one sees "Cheap food is not good" one tends to think of cheap bad food.

Back to jokes: A man came into a restaurant very tired. The waiter came over and said: "What would you like, sir?" The man replied: "Bring me a plate of soup and a kind word!" The waiter went into the kitchen, came back and put a plate of soup on the table and said: "Don't eat it!"

My favorite restaurant joke concerns three friends at a restaurant. The waiter came over to take the order. The first friend said: "I'd like a glass of tea." The second one said: "I want a glass of tea and I want the tea to be hot!" The third said: "I'd like a glass of tea, I want the tea to be hot, and I want the glass to be clean!" "Very good!" said the waiter, who went into the kitchen. Later he returned with three glasses of tea on a tray and said: "Let's see now, which one of you wanted the clean glass?"

There is a good oldie, which I think should be revived: A man in a restaurant said to the waiter: "I'd like a cup of coffee without cream." The waiter went into the kitchen and came back and said: "I'm sorry, sir, we don't have any cream. I can let you have coffee without milk."

A man in a restaurant called over the waiter and said: "I want you to taste the soup!" The waiter replied: "Really, sir, I can't do that, but we will be happy to substitute anything you like." The man kept insisting that the waiter taste the soup, and the waiter then said: "I better call the head waiter." The head waiter came over and explained why it was inappropriate for him to taste the soup. The man kept insisting, and finally the head waiter said: "Alright, where is the spoon?" The man said, "Aha! Aha!"

A man in a hotel came down to the front desk and said to the manager: "I want to change my room!" "But, sir," said the manager, "you have the best room in the hotel!" "Look, I want to change my room." "But I can't give you any better room than the one you have!" "Look, don't argue with me; I want to change my room!" "Alright, if you insist—let's see now, oh yes, Room 605 is a good one and available. Please sign here." After the man signed, the manager asked: "Now that you've changed your room, will you please tell me what there was about the room you had you disliked?" The man replied: "It's on fire."

Back to logic: I like the following characterization of logic by Tweedledee in Lewis Carroll's *Through the Looking Glass*. First Tweedledum said to Alice: "I know what you're thinking about, but it isn't so, nohow." Then Tweedledee said: "Contrariwise, if it was so, it might be; and if it were so, it would be; but as it isn't, it ain't. That's logic."

An especially delightful definition of logic is by Ambrose Bierce in his incomparable *Devil's Dictionary*:

> Logic is the art of thinking and reasoning in strict accordance with the limitations and incapacities of the human misunderstanding. The basis of logic is the syllogism, consisting of a major and a minor premise and a conclusion—thus:
>
> *Major Premise:* Sixty men can do a piece of work sixty times as quickly as one man.
> *Minor Premise:* One man can dig a posthole in sixty seconds; therefore—
> *Conclusion:* Sixty men can dig a posthole in one second.

Speaking of syllogisms, I like the following one:

> Some cars rattle; my car is some car, so no wonder my car rattles!

Incredible as it may seem, a man once actually told me that cats make syllogisms! "Oh yes," he said, "I once saw a cat make a syllogism. The cat jumped over a fence, and so the cat thought: 'Cats can jump over fences. I'm a cat, therefore I can jump over fences.' " When I asked the man how he knew that the cat thought that, he was unable to answer.

Is the following syllogism valid?

Major Premise: Everyone loves my baby.
Minor Premise: My baby loves only me.
Conclusion: I am my own baby.

Well, is that valid? (Answer later on.)

Coming back to Ambrose Bierce, his *Devil's Dictionary* is a book I highly recommend. It is full of delightfully humorous definitions. Here are some choice samples:

EGOTIST: A person of low taste, more interested in himself than in me.

LAWYER: One skilled in circumvention of the law.

AMBIDEXTROUS: Able to pick with equal skill a right-hand pocket or a left.

ACCUSE: To affirm another's guilt or unworth; most commonly as a justification of ourselves for having wronged him. (This shows good psychological insight, as does the next.)

JEALOUS: Unduly concerned about the preservation of that which can be lost only if not worth keeping.

Remarks: This is, if anything, even more insightful than the previous definition. An equally insightful and interesting observation was made by Miguel Cervantes in the second volume of *Don Quixote*; he says about envy: "That worm of all vices! The only vice in which no pleasure is attached."

To continue with the *Devil's Dictionary*:

> ASPERSE: Maliciously to ascribe to another vicious actions which one has not had the temptation and opportunity to commit.
>
> DEFAME: To lie about another. To tell the truth about another.
>
> DECIDE: To succumb to the preponderance of one set of influences over another set.
>
> GRAVITATION: The tendency of all bodies to approach one another with a strength proportional to the quantity of matter they contain—the quantity of matter they contain being ascertained by the strength of their tendency to approach one another.
>
> NIHILIST: A Russian who denies the existence of anything but Tolstoi. The leader of the school is Tolstoi.
>
> OATH: In law, a solemn appeal to the Deity, made binding upon the conscience by a penalty for perjury.
>
> THEOSOPHY: An ancient faith having all the certitude of religion and all the mystery of science.
>
> INCOMPOSSIBLE: Unable to exist if something else exists. Two things are incompossible when the world of being has scope enough for one of them, but not enough for both—as Walt Whitman's poetry and God's mercy to man.

Remarks: This last definition reminds me of Mark Twain's definition of a good library: A good library doesn't necessarily have to have any books; all that is required is that it has no books by Jane Austen. I will return to Mark Twain shortly.

Incidentally, I happen to love Walt Whitman! I once went into a bookshop in New York City and asked the proprietor whether she had any material on Walt Whitman. She replied: "As a matter of fact, *you* remind me of Walt Whitman. I guess people have told you that before?" I answered: "Nobody has ever told me that I remind *you* of Walt Whitman!"

Speaking of Whitman, I must tell you the following delightful story: A college girl once told her father that she wanted to visit Walt Whitman. The father angrily said: "No daughter of mine is going to visit such a dissolute character!" However, the girl was adamant, and the father, seeing that she had a will of her own, finally said: "Well alright, but your brother and I are going to go with you!" She said "Fine." And so the three, not living far from Whitman, took their horse and carriage and went over to Whitman's, who received them courteously. The two children engaged in a conversation with Whitman and the father was at first sullen and silent. Gradually he warmed up to Whitman and finally said to him: "Why don't you come home with us and have dinner?" Whitman agreed, came over with them and stayed at their house for three months.

That's what I call a reasonable father!

Answer to the "Everybody loves my baby" syllogism: Before giving the answer, I must emphasize that there is a vital distinction between a syllogism being *valid* and being *sound*. An argument is called *valid* if the conclusion logically follows from the premises, regardless of whether the premises are themselves true or not. An argument is called *sound* if it is not only valid, but also the premises are true (and hence, so is the conclusion). For example, the following argument is not only valid, but sound: All men are mortal, Socrates is a man, therefore Socrates is mortal. In this argument, both premises are true, and the conclusion is a logical consequence of them. By contrast, the following argument, though clearly not sound, *is* valid! All men live under water, Socrates is a man, therefore Socrates lives under water. Obviously the first premise is false, but the conclusion is nevertheless a logical consequence of the premises (in the sense that if all men *did* live under water, then Socrates, being a man, would be one of those who live under water).

Now, the "Everyone loves my baby" syllogism, though obviously not sound, is valid, surprising as it may at first seem! The reason is this: Assume that everyone loves my baby and my baby loves only me. Since everyone loves my baby, then, in particular, my baby loves my baby, and since I am the *only* one my baby loves, then I and my baby must be the same person.

Coming back to Mark Twain, do you know his definition of the German language? He defines the German language as that language in which all the verbs come in the second volume.

When he was in Germany, he once went to the opera, and later reported: "I enjoyed it too, in spite of the music."

About the music of Richard Wagner, he said: "It's probably not as bad as it sounds."

Speaking of Wagner, I once saw a modern play which all took place in Hell. A visitor was very surprised to hear from the Devil that Wagner was in hell. "Of course," said the Devil, "he was a vicious anti-Semite!" "Oh," said the visitor, "but he wrote such beautiful music!" "Ah!" said the Devil, "his music went to Heaven. *He* went to Hell!"

Returning to Mark Twain, he was once at a banquet and very tired. When the time came for him to make a speech, he got up slowly and wearily and said: "Homer is dead, Shakespeare is dead, and I am none too well."

My favorite Twain story is the one in which he was giving a presentation in Maine or Vermont, but couldn't get a rise out of anyone! He made his jokes funnier and funnier, but to a totally dead-pan audience; he couldn't get so much as a smile. He was wondering, "Am I losing my touch?" Well, during intermission he overheard an elderly couple discussing his act, and the man said to his wife: "Weren't he funny? Weren't he funny? You know, at times I could hardly keep from laughin'!"

The essence of Vermont humor, it seems to me, is the fact that when you ask a Vermonter a question, the answer you get, though correct, is totally inadequate! For example, one Vermont farmer went over and asked his farmer neighbor: "Lem, what did you give your horse that time when he had the colic?" Lem answered "Bran and molasses." The farmer went away and returned a week later. "Lem, I gave my horse bran and molasses and it died!" Lem: "So did mine."

A man was once motoring in Vermont. He came to a fork in the road and one sign said: To White River Junction. The other sign also said: To White River Junction. He saw a Vermonter standing by, and so he asked him: "Does it make any difference which of these two roads I take?" The Vermonter replied: "Not to me it doesn't."

The following incident is true: A former student of mine told me that he was once motoring through Vermont and passed a farm house and saw the farmer on the porch rocking himself in a rocking-chair. He asked the farmer: "Have you been rocking like that all your life?" The farmer replied: "Not yet."

The famous Vermonter Calvin Coolidge (called "Silent Cal") was once at a banquet, sitting next to a girl. A half hour passed, and Coolidge said not a word. The girl finally said, "Mr. President, I have a bet that I can get more than two words out of you!" Coolidge replied, "You lose!" (That certainly sounds like creative intelligence, doesn't it?)

On another occasion, Coolidge returned from church and his friend enquired what the preacher had spoken about. Coolidge replied: "Sin." His friend then asked: "What did he have to say about it?" Coolidge replied: "He was against it."

There is one story I read about Coolidge which increased my respect for him enormously! He once came into his office and found a burglar rummaging through his things. When asked what he was doing, he told Coolidge that he was desperately looking for money to go to his dying mother in another city. Coolidge handed him some money and told him to pay him back as soon as he could, and to be careful escaping from the building, since it was heavily surrounded.

There is the story told about Coolidge visiting a farm with some friends. One of them looked at the sheep and said: "Those sheep have just been shorn." Coolidge replied: "Looks like it from this side."

A similar story concerns a mathematician and a physicist who took a plane from San Francisco to Washington D.C. After arriving in Washington, they were supposed to write a report of anything interesting they saw from the plane. Well, at one point they saw a black sheep in Kansas. When they arrived in Washington, the physicist wrote: "There is a black sheep in Kansas." The mathematician wrote: "There exists, somewhere in the Midwest, a sheep, black on top."

Do you know the difference between a physicist and a mathematician? The following is a good test to determine which type one is: We have a cabin in the woods with an unlighted gas stove, a book of matches, an empty pot and a faucet of cold running water. How would you go about getting a pot of hot water? Just about everyone answers that they would pour cold water from the faucet into the pot, put the pot on the stove, and light the stove. I then explain that so far, mathematicians and physicists are in agreement. But now comes the crucial test: this time the conditions are the same as before, except that now you have a pot already filled with cold water. How would you now get a pot of hot water? The usual reply is to put the pot of cold water on the stove and then light the stove. I then explain: "Then you are a physicist! A mathematician would dump out the water, reducing the problem to the preceding one, which has already been solved."

A more drastic version is this: You have a fire hydrant, a hose and a building on fire. How do you put out the fire? Obviously you connect the hose to the hydrant, turn on the water and put out the fire. But now, we have instead the following: A fire hydrant, a hose, and a building not on fire. How do you put out the fire? The physicist does nothing, whereas the mathematician sets the building on fire, reducing the problem to the preceding one, which has already been solved.

To interrupt this train of thought for a moment, do you know how to tell whether a given bird is male or female? Well, you offer it some bird seed. If he eats it, it's male; it she eats it, it's female.

"Yes," you might say, "but how can you tell if it's a he or a she?" The answer is simple. If it's a male, then it's definitely a he, if it's a female, then it's a she. Thus, you should have no difficulty!

This is reminiscent of Lewis Carroll's question of which is better, a clock that loses a minute a day, or a clock that doesn't go at all? Most people would opt for the one that loses a minute a day, but, as Lewis Carroll pointed out, the clock that loses a minute a day is right only once every two years, whereas the other clock is right twice a day. "Yes," you say, "but what's the use of its being right twice a day, if I can't tell when the time comes?" Well, let's suppose the clock points to eight o'clock. Then when eight o'clock comes around, the clock is right. "But," you say, "how am I to know when it *is* eight o'clock?" Ah, now; this is the subtle part. You keep your eye *very* carefully on the clock, and the instant it is right—then it will be exactly eight o'clock!

This reminds me of a lovely problem: A certain man had forgotten to wind his clock and had no accurate idea of what time it was. He had a dinner invitation for the evening with a friend, to whose house he planned to walk. He did not know how long the walk would take. Well, he had the following ingenious idea for setting his clock right. He wound it up, and then set it at 12:00. He then went to his friend's house, noticing the time he arrived and the time he left. When returning home, he looked to see what time the clock read, and he could then deduce the correct time. How? We assume that the path to his friend's house was level and that the time it took to walk there was the same as the time it took to walk back. (Solution is given following the next item.)

Coming back to physicists and mathematicians, do you know what characterizes mathematicians, physicists, and engineers? This is nicely illustrated by the story of a mathematician, physicist, and engineer who were professors at a university and had adjacent offices. One day, each one of them smoked his pipe in his office, dumped his hot ashes in the waste basket which had paper, and all three baskets caught fire. Well, the engineer took out his slide rule and calculated approximately how much water was necessary to quench it, took approximately that amount and successfully quenched the fire. The physicist computed the upper and lower limits necessary, took the average, and quenched the fire. The mathematician, using far more sophisticated and refined techniques, computed *exactly* how much water was necessary, and went back to work.

Solution to the time problem: Let's say he arrived at his friend's house at 7:00 and left at 10:00, and when he got back his clock read 4:00. Then he was gone 4 hours, stayed with his friend for 3 hours, and he walked 1 hour, a half hour each way. Since he left his friend's house at 10:00, it must now be 10:30! (Ingenious, eh what?)

There is the joke told about the famous mathematician John von Neumann that he was called in as a consultant for a project to send up the first rocket ship, which was already partly built. After looking it over, he asked, "Where did you get the plans?" The project manager replied: "We have our own staff of engineers!" "Engineers?" asked von Neumann, "why the entire theory of rocketry can be found in my 1952 paper!" Well, they scrapped the five million dollar structure and rebuilt the rocket exactly according to von Neumann's plans, but when they started it, the whole thing blew up! They angrily called him back and told him: "We built it *exactly* according to your plans, and the minute it started, it blew up!" "Oh, yes!" said von Neumann, "that's known as the blow-up problem. I treated that in my paper of 1954."

There is a joke told about some people who got lost hiking in the mountains. They decided to test the echo, and one yelled at the mountain, "WE ARE LOST!" Soon after a voice came back: "YOU ARE LOST!" "Well," said one, "that was obviously not an echo, but a person." "Yes," said another, "and a mathematician." "How do you know he was a mathematician?" "Obvious, for three reasons. First, he took a long time answering. Secondly, the answer was absolutely correct, and third, totally useless."

A teacher once said to a grade school kid: "If your father had ten dollars and you asked him for six, how many would he have left?" The kid replied: "Ten." The teacher said: "You don't know your math!" The kid replied: "You don't know my father!"

One of my favorite math jokes is about a physicist who came to the office of a mathematician friend and said: "I just concluded an experiment that conclusively proves that quantity A is bigger than quantity B." The mathematician replied: "That is perfectly understandable. You didn't even have to make the experiment; A must be bigger than B for the following reasons...." The physicist interrupted and said: "Oh, I made a mistake, I meant to say that it is B that is bigger than A." The mathematician replied: "That is even more understandable; here is why...."

Mathematicians are notoriously absent-minded. One mathematics professor met a student in the hall one afternoon, who asked him whether he had lunch. The professor paused for a moment and said: "Which way was I walking when I met you?"

On another occasion, a mathematician was at a mathematics convention. He forgot what his car looked like, and hence which car in the parking lot was his, and so he waited until all the other cars were gone, and took the remaining one.

The most extreme case I know is of a mathematician teaching at Harvard, whom I will call Professor *Greylock*. At one point he and his wife moved from one part of Cambridge to another. His wife, knowing his absent-mindedness, decided to condition him 30 days in advance, and so said to him: "Now, in 30 days we will have moved, and so when you get out of class, you don't take bus A, you take bus B!" He replied: "Yes, dear." The next day she said: "Now remember, in 29 days we will have moved, so you don't take bus A; you take bus B." "Yes, dear." This went on day after day, until the final moving day came, and so in the morning she said: "Now today is the day we move, and so when you get out of class, you take bus B, not bus A." "Yes, dear."

Well, after class he took the usual bus A, and when he got to his house, it was empty, and he said to himself, "Oh, of course! This is the day we moved," and so he walked back to the square, took bus B and got off at the right stop, but forgot his address! It was getting dark, and he was quite near-sighted, and after groping around for a while, he saw a little girl on the street and asked her: "Do you by any chance happen to know where the Greylocks live?" She replied: "Oh, come on, Daddy, I'll take you home."

This same professor, during a lecture, said at one point: "This is obvious." One student raised his hand and asked: "Why is it obvious?" The professor replied: "Er, just a moment," and walked out of class and went into the hall, pacing back and forth, and returned to class 15 minutes later, went to the blackboard and said: "Yes, it is obvious!" and continued with the lecture.

When I was a graduate student at Princeton, the following joke went around about the word "obvious": when Professor A. says something is obvious, it means that if you go home and think about it for a week, you will realize it is true. When Professor B. says something is obvious, it means that if you go home and think about it for the rest of your life, the day *might* come that you realize it is true. When Professor C. says something is obvious, it means that the class has already known it two weeks ago. And when Professor D. says something is obvious, it means it's false!

It interested me to hear one mathematician observe that in mathematical papers which contain an error, the most likely place for the error to occur is when something is said to be obvious. One usually says that something is obvious when he or she doesn't wish to bother giving the details, and that is just when mistakes are likely to slip in.

Let us suppose that we are told that in a certain club, all Frenchman in the club wear hats. Later we find out that there are no Frenchman in the club. Then is this statement that all Frenchman in the club wear hats to be taken as true, false, or inapplicable? People's opinions differ on this, but in the kind of logic used in so-called *classical* mathematics, and also in computer science, the statement is taken to be true! The only way the statement could be falsified is by finding at least one Frenchman in the club who doesn't wear a hat. Thus the statement is taken to mean nothing more nor less than that there is no Frenchman in the club who doesn't wear a hat, and so if there are no Frenchman in the club at all, then there are certainly no Frenchmen in the club who *don't* wear hats.

This is an illustration of a basic fact used by classical logicians and computer scientists: a false proposition implies *every* proposition! Here is another illustration: Suppose I put a card face down on the table and say: "If this card is the queen of spades, then it is black." You would certainly agree. Then I turn over the card and you see it is red—say, the seven of diamonds. Would you then take back your agreement with my statement? In logic and computer science, the statement would still be regarded as true. The antecedent "if this card is the queen of spades" of the statement is now seen to be false, and a false proposition is said to imply *every* proposition. This fact seems to go counter to many people's common sense, which is quite understandable. Indeed, there is a field called *relevance logic*, in which it is *not* the case that a false proposition implies everything. But in classical logic, the fact holds. There is the true story of a man, who couldn't accept this fact, who said to Bertrand Russell: "You mean to say that from a false proposition, one can infer *anything*?" Russell replied "Yes." "Really?" said the man, "If, for example, we started with the assumption that two plus two is five, could you then prove that you're the Pope?" Russell, with his typical sense of humor, replied: "Quite easily! Let's assume that $2 + 2 = 5$. Subtract 2 from both sides, we get $2 = 3$. Now subtracting 1 from both sides we get $1 = 2$. Transposing, we get $2 = 1$. Now, the Pope and I are two. Since two equals one, then the Pope and I are one. Thus I am the Pope."

Well, when I was a graduate student in Princeton, I gave Russell's proof to my future wife Blanche. After having shown her that starting with $2+2 = 5$, I can prove that I'm the Pope, I then told her that from $2 + 2 = 5$, I can also prove that I'm *not* the Pope. She replied, "Is that how you spend your days in Princeton? One day he proves he *is* the Pope; the next day he proves he's *not* the Pope."

Incidentally, I wondered why the one who posed the question to Russell ever thought of Russell deducing that he was, of all things, the Pope! I suspect that he was unconsciously influenced by the following verse of Lewis Carroll: "He thought he saw an argument that proved he was the Pope. He looked again and saw it was a bar of mottled soap."

Someone once said about a certain composer: "He wrote only one composition, and that was one too many!"

That is similar to a remark written by Samuel Johnson to an author who had sent him a manuscript: "Your manuscript is both good and original. Unfortunately the good parts are not original, and the original parts are not good."

This is turn is similar to an incident in which someone once sent a manuscript to Disraeli (when he was Prime Minister of England). Disraeli had no interest in the manuscript and wrote back to the author: "I can assure you, Sir, that I will lose no time in reading it."

Disraeli really had quite a wit! In a parliamentary debate between Disraeli and Gladstone, at one point, Gladstone said: "If you continue in this way, you will either end up on the gallows, or die of a social disease." To which Disraeli replied: "That depends on whether I embrace your principles or your mistress!"

Ah, the way people spoke in the eighteenth and nineteenth centuries! I recall an incident about Benjamin Franklin: before the Revolutionary War started, there was some talk of a possible compromise. Franklin went with some friends to the house of the British General Howe to discuss matters. At one point Howe said: "I can't tell you how much it would pain me to have to ravage and destroy America!" At which point Franklin smiled and said: "I can assure your Lordship that we will do our utmost endeavors to spare you this mortification!"

I am reminded of the fact that when I was a child, my brother Emile, ten years older than I, gave me the following delightful definition of a gentleman: "A gentleman is one who does not hurt other's feelings unintentionally."

Coming back to Samuel Johnson, he was once in a restaurant and ordered coffee and cream. The waitress said: "Alright, coffee with cream." Johnson replied: "No, not coffee *with* cream; coffee *and* cream!" The waitress said: "What's the difference?" Johnson replied: "Madam, don't you know the difference between a mother *and* child and a mother *with* child?"

On another occasion, Johnson was in a stage coach and hadn't bathed for a while. A woman sitting opposite of him said: "You smell, Sir!" Johnson replied: "On the contrary, it is *you* who smell; it is *I* who stink!"

This bears a similarity to the incident in which the wife of an English professor came home and found her husband with another woman. She said: "I *am* surprised!" He replied: "No, my dear; it is I who am surprised. You are astonished!"

Samuel Johnson was the most amazing combination of one who at times had remarkable psychological insight, and at other times had the most ridiculous narrow-minded prejudice! As an example of the latter, he was once discussing America with Boswell, and said about the Americans: "Oh, they are just a race of convicts who deserve anything we send them short of hanging!"

On the positive side, as an example of his excellent psychological insight, in one of his articles he wrote on why people tend to reveal secrets entrusted to them; he said: "The vanity of having been trusted with a secret is one of the main motives for betraying it."

Now I cross the ocean and a century or so and turn to Abraham Lincoln. He was once approached by the president of a railroad company who tried to sell him some stock in his railroad. He described the various good features of his railroad and at one point said: "Moreover in my railroad, a collision between trains is impossible!" "Maybe highly improbable, but certainly not *impossible*!" said Lincoln. "No!" he replied, "I say impossible!"

Why did he say *impossible* instead of *improbable*? (Answer given later.)

On another occasion, a certain author visited President Lincoln and tried to sell him one of his books. Lincoln was not interested. The author then said: "Well, Mr. President, since you don't want the books for yourself, could you at least write an endorsement for it, to make it easier for me to sell it to others?" Lincoln replied: "Certainly," and wrote the following endorsement: "Those who like this kind of book, will find it just the kind of book they like."

This reminds me of an advertisement I once saw for a weight-reducing program, which read "Guaranteed to lose up to five pounds a week." A pretty safe guarantee, eh?

Solution to the last problem: The railroad system had only one train.

I once played a joke on one of my Ph.D. students who was in the process of writing his dissertation. One summer he was away, and I wrote him a letter which I ended with: "And if you have any questions, don't hesitate to call me collect and reverse the charges." (Get it?)

I once saw a cartoon in which a salesman, in the process of trying to sell a certain appliance, said: "What's more, this sells for less than those costing twice as much!"

Someone once asked me: "Have you ever heard of caution children?" "No," I replied, "what I the world are caution children?" He replied: "I don't know. All I can tell you is that the other day I was driving along and I saw a sign on the road: *CAUTION CHILDREN*."

Why is it difficult for Egyptian mummies to make friends? Answer: they are too wrapped up in themselves.

One man told a friend that he had the largest sheep farm in the state. When asked how many sheep he had, he said: "I don't know. Every time I start counting them, I fall asleep."

Why is it logically impossible for there to be more than one dock in the universe? Because if there is more than one, you would have a pair-o-docks.

Do you know the difference between Heaven and Hell? Heaven is where you find British policemen, a French chef, German mechanics, Italian lovers and Swiss organizers. Hell is where you have German policemen, a British chef, French mechanics, Swiss lovers and Italian organizers.

I recently heard the following double blonde joke: A blonde driver was speeding and was pulled over by a blonde police lady who asked her for her driver's license. She looked in her purse and pulled out a mirror, looked into it and seeing herself, thought it was her license, and handed it to the police lady, who looked into it and said: "If I had known you were a cop, I wouldn't have pulled you over!"

Another: A man was driving and was pulled over by a policeman who said: "This is a forty-five mile zone, and you were going at fifty-five miles an hour!" The man said: "No, I was going at forty-five." The cop said: "No, you were going at fifty-five!" The man said: "Well maybe I was going at forty-eight." The cop said: "No, you were going fifty-five," and so they argued back and forth, until the wife turned to the officer and said: "No sense arguing with him, officer; he's had too much to drink."

A novice at an office had a manuscript in his hand and was standing by a shredding machine. He didn't know how it worked, and asked one of the experienced girls there to help. She put the manuscript in the machine and the man said: "Fine, but where do the copies come out?"

Another police joke: About seventy years ago, a man was pulled over by a cop for speeding. The cop took out his pencil and notebook and asked him: "What's your name?" The man replied: "Spiliopoulas Evangopelous Papperilosopoudos." The policeman said: "Well don't let it happen again!"

A very pompous man visited a mental institution and got into an argument with one of the patients. At one point he drew himself up very stiffly and said: "Do you know who I am?" The patient said: "No, but they can tell you at the information desk."

A man told his friend that he had not spoken to his wife for sixteen days. When the friend asked him why, he said: "I didn't want to interrupt."

A high school student was taking an exam. At one point the proctor said: "Examination is closed. Stop writing!" The student wrote for 30 seconds more, and when he took his paper to the desk, all the other papers had already been handed in. The proctor said: "I can't accept your paper. You cheated. You wrote overtime!" The student said: "Do you know who I am?" The proctor said: "No." The student said: "Good!" and thrust his paper into the stack of papers and quickly walked away.

A college freshman asked his advisor what to study. The following conversation took place:

ADVISOR: You should study logic.
STUDENT: What is that?
A.: Logic enables one to infer one proposition from another.
S.: Can you give me an example?
A.: Yes. Do you have a lawn mower?
S.: Yes.
A.: From which I infer you have a lawn.
S.: Yes, I do.
A.: Then you must have a house.
S.: Yes.
A.: And you are probably married.
S.: Yes, I am.
A.: And you have children.
S.: Yes, I have two.
A.: From which I infer that you are a heterosexual male.
S.: *(laughing)* I am indeed a heterosexual male. Gee, this logic is amazing! From the fact that I

> have a lawn mower, you could deduce that I am
> a heterosexual male! This is really amazing!

Later the student went into the hall and met a fellow freshman. He said to him: "You should really study logic!" The other asked: "What is that?" He replied: "Logic enables one to deduce one proposition from another. For example, do you have a lawn mower?" The other replied, "No." He said: "You faggot!"

Two gay men were standing by a fence. A married couple passed by arguing violently. One gay man said to the other: "See, I told you these mixed marriages never work!"

A man went to a priest and said: "I'm ninety years old and yesterday I made love to three women." The priest said: "Really? When did you last go to confession?" The man replied: "I don't go to confession; I'm not Catholic." The priest said: "Then why are you telling this to me?" The man replied: "I'm telling it to everybody!"

I saw a French movie in which two old men were standing by a fence. A beautiful girl walked by, and one of the men sighed and said to the other: "At times like this, I wish I were eighty again!"

Getting back to logic, my introduction to logic was at the age of six. It happened this way: I was in bed with a cold or grippe, or something, and my brother Emile came to me in the morning and said: "Today is April Fool, and I will fool you like you have never been fooled before!" And so I waited all day long for him to fool me, but he didn't. When night-time came, my mother asked me why I didn't go to sleep. I said that I was waiting for Emile to fool me. My mother said to Emile, who was standing by: "Emile, why don't you fool the child?" The following conversation then took place:

EMILE: So you expected me to fool you, eh?
RAYMOND: Yes.
EMILE: But I didn't, did I?
RAYMOND: No.
EMILE: But you expected me to, didn't you?
RAYMOND: Yes.
EMILE: So I fooled you, didn't I?

Well, I spent a good time in bed wondering whether I had really been fooled or not. One the one hand, if I was not fooled, then since I expected to be and wasn't, I was then fooled. But then again, if I was fooled, then I did get what I expected, so in what sense was I fooled?

This is somewhat reminiscent of a short piece I wrote in my book *This Book Needs No Title* titled "Should One Worry?" which goes as follows:

If one worries a lot, one is obviously unhappy, since worry itself is one of the most painful things in life. If one fails to worry enough, then (at least so I have been told) one may be even worse off, because one may fail to take the precautions necessary to ward off even greater catastrophes than worry.

Who is really better off, the happy-go-lucky who enjoys himself from day to day and lets tomorrow take care of itself, or the worrying, prudential person who takes all conceivable precautions for the future, but is always worrying that he is not taking enough precautions?

All my life, people have told me that my main trouble is that I do not worry enough, and I must admit, this thought has always worried me!

Coming back to my childhood, I was born in Far Rockaway, Long Island. When I was an infant of only a few months old, my mother recognized my musicality because when I was wheeled outside in my baby carriage, and I would hear a bird sing, I sang back the very same note.

A very revealing incident occurred when I was about two or three years old. I used to sit on my grandfather's lap and liked to play with the smoke from his cigarette. On this occasion, he was not smoking, and I wanted him to smoke, and kept saying: "Moke! Moke!" He ignored it, and I repeated: "Moke! Moke!" This occurred several times, and so to distract me, he told me a long, long story. I sat quietly and patiently, listening intently. But when at long last the story was over, I said "Moke! Moke!"

My grandfather smiled, and said to my mother, "This is a good indication of how he will be when he grows up!" As a matter of fact, my friends, when they hear this story, tell me that I haven't changed!

Speaking of grandparents, do you know why they get along so well with their grandchildren? Answer: Because they have a common enemy.

A riddle I thought of: I know a man who has great grandchildren, yet none of his grandchildren have any children. How is this possible? (Answer given later.)

Coming back to my own life, in my childhood, I studied both piano and violin. My interest at the time, though, was not music, but science—particularly chemistry. I had my own chemical laboratory in a third floor little used bathroom. I also built radios at the time, and also I connected a long wire from my house to the next door house where my friend Bernard lived, and we communicated to each other by Morse code. When I was twelve years old, I entered a city-wide piano contest, but did not win the gold medal (which was the first prize), but only a silver medal (which several people won). The next year I entered again, and this time won the gold medal. The day before the contest, a funny thing happened. I got up early before my parents awoke, and together with my friend Bernard, we carried a canoe we had built down to the ocean. Unfortunately we did not know that we were supposed to put lead on the keel to balance the boat, and so it toppled over and dumped us both into the water! Later in the day, when my parents heard about this, they were appalled at my doing this so soon before the contest! But this did not stop me from winning.

At the age of thirteen, we moved to the city. I continued with piano lessons, but not violin lessons, though I still played the violin in the high school orchestra. When I studied geometry, a whole new world opened to me! It was then that I fell in love with mathematics, and couldn't decide whether to become a mathematician or a concert pianist, but hoped I could do both. The only courses I had in high school that were of any value to me were chemistry, physics and geometry. I then decided I had enough of school, and wanted to study on my own, and so dropped out of high school. For several years following, I referred to myself as a "born-again dropout." In these days I neither went to school nor had a job, and was generally regarded as a general good-for-nothing idler. I recall that at a party I once attended, someone asked me what I was doing these days. I replied: "I'm waiting for the meek to inherit the earth."

Actually I was not nearly as idle those days as people around me thought. For one thing, at the time, I learned calculus on my own, as well as group theory and elements of the field known as Galois theory. Secondly, it was in those days that I invented scores of chess problems, which I did not publish until many, many years later, in two volumes: *The Chess Mysteries of Sherlock Holmes* and *The Chess Mysteries of the Arabian Knights*. (The word *knights* is not misspelled; it really is in the title.) How these books ever got written and particularly how they got published, is quite a story in itself, which I will now tell you.

Leaving my high school days (which I will come back to later on), skipping several years ahead, when I was a graduate student at Princeton University, I showed one of my best chess problems to my office mate, as well as to a distinguished logician who was then a Fellow of the nearby Institute for Advanced Study. My office mate said to me, "Why don't you publish this unusual problem before someone else does?" I laughed and said: "Why would anyone want to do that?"

Well, several weeks later, the logician from the Institute met me and said: "How come someone published your problem in the Manchester Guardian without crediting you as the author?" (The Manchester Guardian is a British newspaper.) I then got hold of a copy of the paper and saw to my amazement that there was my problem submitted by none other than the father of my office mate!

I immediately went to my office mate and asked for an explanation. "Oh yes," he said, "I showed your problem to my father who has been in frequent correspondence with the chess editor of the Manchester Guardian, and he sent him your problem with a suggestion that he publish this unusual problem instead of the usual type of chess problem."

The "unusualness," by the way, is that this problem, like most of my chess problems, is not of the usual type—White to play and win in so many moves—but is of the type called "Retrograde Analysis" and are really studies in chess logic. That is, a position is given, and from it the reader must deduce what happened in the past of the game. These problems have very much the character of detective stories.

Now, the father who submitted the problem never claimed to be its author, but failed to say who the author was. When I expressed my disappointment at not being credited, my office mate told me that he would speak to his father about this. A couple of weeks later, I received a very nice letter from the chess editor of the Manchester Guardian expressing regret that he had not known who was the author of this "delightful work" and that my authorship would be acknowledged in the next issue. He also asked if I had any other problems of a similar nature, and that if I had, I should send them to him to look at. And so for the next several months some of my problems got published in the Manchester Guardian.

Then, equally curious, some years later, a remarkable thing happened: the very same problem which the father of my office mate sent to the Manchester Guardian—this problem appeared in Martin Gardner's puzzle column of the Scientific American magazine, with the remark: "Author Unknown." It had been sent in by a reader with a note saying that while he found it remarkable, he did not know who had originated it. I knew nothing about this at the time, and probably would never have known if it were not for the fact that twenty years earlier, I had shown the problem to a fellow mathematics student, now Professor Mitch Taibleson, who saw my problem in the Scientific American, and promptly wrote to Martin Gardner that the problem had been devised earlier by Raymond Smullyan, and that it was one of a large collection of unpublished chess problems invented by Smullyan when he and I were fellow students at the University of Chicago.

If it hadn't been for that letter, my whole life might have turned out differently! But as it happened, the letter led to a happy renewal of my acquaintance with Martin Gardner, who wrote to me about the matter and urged me to stop dallying and get the book written. This got me to go to work.

I had originally planned to incorporate my retrograde chess problems into Arabian Nights stories in which the chess pieces would represent people—the White King would be Haroun El Rashid, one of the White bishops would be his grand vizier, etc.—but then the well-known retrograde expert Mannis Charosh, who had seen some of my problems, sent me an excellent article he had written on the subject titled "Detective at the Chessboard." This title immediately captured my fancy, and I thought, "Why not have an *actual* detective at the chessboard, and for that matter why not Sherlock Holmes?" And so I changed my original plan and decided to divide my hundred best chess problems into two groups of fifty, and then wrote one book titled "The Chess Mysteries of Sherlock Holmes," in which I encapsulated each problem into a detective story, and the other which I decided to title "Chess Mysteries of the Arabian Knights" (instead of "The Chess Mysteries of the Arabian Nights").

To digress for a moment, speaking of Sherlock Holmes, do you know the story about Holmes and Watson out camping? At about 2:00 A.M., Holmes wakes Watson and says: "Look straight above you, and tell me what you can deduce." Watson looks up and says: "Well, I see all these beautiful stars up there, from which I infer that there might be life elsewhere in the universe." Holmes replied: "No, no, Watson. What you *should* have deduced is that someone has stolen our tent!"

Coming back to my story, after having written the two books, the next problem was how to get them published. Well, I sent copies of the manuscript to Martin Gardner, who forwarded them to a well-known publishing house. The editor to whom he sent them phoned me and was very enthusiastic about both of them. However, soon after, he sent me a short letter telling me that although he thought very highly of the books, the sales department turned them down! As he said, "Times have changed and these days publishers want books with good commercial possibilities." So here I was, stuck with what I thought were two excellent manuscripts.

About that time, my colleague Professor Melvin Fitting was asked by his father-in-law Oscar Collier, a literary agent and editor for Prentice Hall, if he would be interested in writing a book of logic puzzles. Mel replied that it was really not in his line, and that Raymond Smullyan might be a good one for that. And so Oscar got in touch with me and I told him about my chess manuscripts. He said that he was not interested in chess problems—he wanted a book of logic puzzles. And so I got to work and wrote my first book of logic puzzles titled *What is the Name of this Book?*

After I wrote a few chapters, I had lunch with Oscar one day and brought the chapters to him and also told him of my chess books. As he told me before, he was not interested in my chess books for Prentice Hall, but would definitely propose my book of logic puzzles, which he did, and which Prentice Hall published. Thus came out my first book of logic puzzles. As for my chess books, Oscar Collier, being a literary agent, told me that he would try and find a publisher for them. He sent my chess books to the house of Alfred A. Knopf, where they were looked at by a senior editor Ann Close. She promptly sent them back to Oscar Collier, telling him that chess was out of her line. But then my *What is the Name of this Book* came out with a rave review from Martin Gardner who pronounced it "the best book of logic puzzles ever written." When Ann Close read this review, she telephoned Oscar Collier and said that she would like to have a second look at my chess manuscripts. Now Ann is a very clever lady who knew virtually nothing about chess at the time, but applied herself so thoroughly that after a month she had enough expertise to not only understand my chess problems, but make corrections to several of them! The books were then published by Alfred A. Knopf.

That is the curious story of how my chess books got published. Alfred A. Knopf subsequently published many more—in fact most of my logic puzzle books. As of now, I have 30 books published, with one more about to come out. My puzzle books are more than just books of recreational puzzles; they are designed to introduce the general reader to deep mathematical and logical results through recreational puzzles.

Returning now to my high school days, as I told you, I dropped out to study mathematics on my own. I never graduated high school, but got into college by taking the College Board entrance exams.

Speaking of school boards, I like what Mark Twain said: "At first God made idiots—that was for practice. Then He made school boards."

One amusing incident: One of the College Board exams was of course in English, in which the applicant asked to write an essay on a novel he had read. Well, I was in a mischievous mood at the time, and simply made up a novel on the spot, and wrote an essay on this nonexistent novel! I passed the exam very well.

I went to several colleges, and kept dropping out and coming back. At the University of Chicago I took courses, mainly in the mathematics department, but the courses that interested me most were in mathematical logic, given not in the mathematics department, but in the philosophy department. (Mathematical logic, which is my field, is taught in mathematics departments, philosophy departments and computer science departments.) My teacher in mathematical logic was the world-famous philosopher and logic professor Rudolph Carnap, who had a major influence on my life.

I supported myself in those days as a close-up magician, working in various night clubs. I went from table to table entertaining customers. At that time I also taught piano at Roosevelt College. This, in fact, was my first teaching job.

Then something remarkable happened. At Dartmouth College in Hanover, New Hampshire, the mathematics department needed an instructor. The then head of the department John Kemeny (who later became president of the college) contacted Rudolph Carnap and asked him if he knew someone who could do the job. On the basis of term papers I had written for Carnap (which later got published), he recommended me quite strongly even though I had not yet got even a bachelor's degree. I was then asked to come up to Dartmouth for an interview; the interview went well and the department was in favor of hiring me, but first this had to be approved by the dean of the college. When Kemeny phoned the dean about the prospect of hiring me, he told him that I didn't yet have a degree. The dean thought that Kemeny meant a doctoral degree. Then Kemeny said: "You don't understand: He doesn't even have a *bachelor's* degree!" There was a pause. Then the dean said: "Well, if you think he is competent, then hire him." And so I got hired for a year. The department was evidently pleased with me, and hired me for a second year. Meanwhile the University of Chicago gave me a bachelor's degree on the basis of courses I never took, but taught. After teaching two years at Dartmouth, I then went to Princeton to work on a Doctor's degree, which I got in three years.

I taught at Princeton for several years. One particular teaching incident is quite funny: I was teaching a course in probability, and I told the class something that surprised them (and will probably surprise those of you who do not know it)—namely, that if there are 24 people in a room, the chances that at least two of them have the same birthday is greater than fifty percent. (With 30 people in a room, the chances are overwhelming!) I then told the class that since there were only nineteen in the room, the chances that two have the same birthday is very small. One boy said: "I'll bet you a quarter that two of us here have the same birthday." I thought about this for a moment and said: "Oh, of course! You know the birthday of someone else here as well as your own!" He replied: "No, I give you my word that I don't know the birthday of anyone here other than my own. Nevertheless I'll bet you that there are two of us here who have the same birthday." I said "OK" and took the bet and started pointing to the students one by one, asking them their birthdays, until at one point I suddenly stopped, realizing that I had forgotten that two of the boys there were identical twins! Boy, did the class have a good laugh!

This only proves the futility of pure theory when not backed by empirical observation!

Speaking of probability, here is a puzzle that most people get wrong.

We have three chests of drawers. Each chest has two drawers. In one chest, each drawer has a gold coin. In one other chest, each drawer has a silver coin. In the remaining chest, one drawer has a gold coin and the other drawer has a silver coin. You pick a chest at random, open one of the drawers and find a gold coin. What is the probability that the other drawer has a gold coin? (Answer given later.)

Again, speaking of probability, there is the story of the statistician who told a friend that he never takes airplanes. When asked why, he replied that he computed the probability that there be a bomb on the plane, and that although the probability was low, it was too high for his comfort.

A week later, the friend met him on a plane and asked him why he changed his theory. He replied: "I didn't change my theory. It's just that I subsequently computed the probability that there simultaneously be two bombs on the plane. This is low enough for my comfort, and so I now carry my own bomb." (This is a typical freshman fallacy!)

Solution to the last puzzle: Many people wrongly believe that the probability is fifty percent. They reason that the chest with the two silver coins is really out of the picture (since a gold coin was found) and that part of the reasoning is correct. But then they reason that the chances are therefore even one picks the chest with the two gold coins or the chest with one gold and one silver. This is correct, before one of the drawers is opened and found to have a gold coin. But after a drawer with a gold coin is found, the story is very different! The proper way to look at the situation is this: As mentioned before, the chest with the two silver coins is quite irrelevant, and so we need deal only with two chests—the one with the two golds and the one with a gold coin and a silver. Thus you are dealing now with four drawers, an open drawer with a gold coin, two closed drawers each with a gold coin, and one closed drawer with a silver coin. Thus there are three drawers left, two of them have gold coins and only one of them has a silver coin. This silver coin has equal probability of being in any one of these other doors. So, regardless of whether you pick the drawer in the same chest, or one of the other two drawers, the chance of finding a gold coin is two to one. Thus the correct answer to the problem is two thirds, not one half!

Three more puzzles:

1. There was a convention of one hundred scientists; each was either a physicist or a chemist. One physicist noted that given any two of the members, at

least one of them was a chemist. From this observation, can it be determined how many of them were chemists and how many were physicists?

2. In a certain flower garden, each flower was either red, yellow or blue. All three colors were actually represented. One statistician observed that whichever three flowers were picked, at least one was bound to be yellow. Another observed that whatever three flowers were picked, at least one was bound to be red. From these two observations, does it logically follow that given any three of the flowers, at least one is bound to be blue?
3. Here is one of my favorite puzzles (one that I did not invent). On the ground floor of a house there are three electric switches—two of them are false and one of them is genuine. The genuine one turns a bulb on the second floor on or off. At the moment, the bulb has been off for quite a while. You are allowed to do anything you want with the switches, and then go upstairs to observe the bulb. You are then to determine which of the three switches is the genuine one.

Answers:

1. To say that given any two of them, at least one is a chemist is but another way of saying that no two are both physicists. Thus there is only one physicist; the other 99 are chemists.
2. The answer (which may surprise some of you) is that there can be only three flowers in the entire garden! Here is why: If there were more than three, then there would have to be at least two of the same color (since there are only three colors). If there were two reds, then one could pick two reds and one blue, thus

avoiding a yellow, which is contrary to the report of the first statistician. If there were two yellows, then one could pick them with one blue, thus avoiding a red, which is contrary to the second statistician's report. If there were two blues, one could pick them with one red, thus avoiding a yellow, contrary to the first report (or alternatively, one could pick them with a yellow, thus avoiding a red, contrary to the second report). Thus there can be only three flowers in the garden. And so whichever three you pick (and there is only one choice) of course one is bound to be blue.

3. At first sight, it might seem impossible that there is a solution, since you can observe only two possibilities—namely, the bulb is either on or off—and that observing only two possibilities cannot distinguish between three possible cases. It is indeed true that observing only two possibilities cannot distinguish three possible cases, but the fact is that there are three possible cases you can observe—namely, the bulb is on; the bulb is off but still hot; or the bulb is off but cold. And so here is one possible solution: Call the three switches A, B, and C. You first turn A on and leave it on for a while. If it is the genuine switch, it will turn the bulb on and heat it. Then you turn A off and put switch B on, leave it on and go upstairs. If the bulb is on and hot, then A is the genuine switch. If the bulb is on but cold you will know that B is the genuine switch. If the bulb is not on, then C is the genuine switch.

Now back to my Princeton days. When I was there I heard that some years before I arrived there was a grade school girl who was doing poorly in arithmetic. After about two months, her work showed a marked improvement. Her mother asked her how it was that her arithmetic improved. She replied: "I heard there is a teacher here who teaches real good. I go to his house every day and he helps me with arithmetic. He teaches real good. I forget his name—it is something like Einstock, or something like that."

Yes, it was indeed Albert Einstein. Every day she came to his house, and of course he helped her.

I am quite fond of one story I heard about Einstein: He once told a colleague that he did not like teaching at a co-ed college. When asked why, he said that with all the pretty girls in the room, the boys wouldn't pay attention to mathematics or physics. The colleague said: "Oh come on Albert, you know that the boys would listen to what *you* have to say." Einstein replied: "Oh, such boys are not worth teaching!"

I spoke earlier of John Kemeny. For some time he was an assistant to Einstein when they were at the Institute for Advanced Study. The great logician (in fact the world's greatest logician) Kurt Gödel was there at the same time. Their offices were opposite each other. At one point Gödel was inventing some imaginary universes in which the laws of physics were different. Kemeny once asked him what Einstein thought of his work. Gödel replied that he didn't know, since he had never been introduced to Einstein. Kemeny was amazed! Here they were at the Institute for several years and never met. Kemeny thought that this was ridiculous, and promptly took Gödel into Einstein's office and introduced them. Since then Einstein and Gödel became the best of friends. Every day after work, the two would be seen walking home together arm in arm.

When Gödel went for his examination for US citizenship, Einstein went with him. On the way, Gödel told Einstein that he would tell the examiner that he found an inconsistency in the US constitution. Einstein replied: "For heaven sakes, don't do that! You will only antagonize them and you will never get your citizenship!" Fortunately, Gödel followed Einstein's advice.

I met Gödel several times, but never met Einstein, since I arrived at Princeton some time after Einstein passed away. My brother Emile—a sociologist and economist—had some interesting correspondence with Einstein and received one particular letter complimenting him on his work on nuclear disarmament. Unfortunately the letter got lost, and was probably stolen.

Let me tell you something amazing! I have a good school book on arithmetic in which Gödel is briefly discussed. There is a photograph under which is the name "Kurt Gödel." The photograph is actually of Einstein. Now, I can understand someone not knowing what Gödel looked like, but I cannot understand that one would not know what Einstein looked like!

When I was a graduate student at Princeton, one particularly interesting event occurred: I frequently visited New York City, and on one of my visits, I met a very charming lady musician. On my first date with her, I asked her whether she would do the following. I'm to make a statement. If the statement is true, would she give me her autograph? She replied: "I don't see why not." I then continued: "But if my statement is false, I want you to agree not to give me your autograph!" She assented. So remember, a true statement gets me an autograph, and a false statement does not—that is very important! Well, the statement I made was such that for her to keep her own agreement she had to give me, not her autograph, but a kiss! In fact the statement I made had to be false (assuming she kept her word) and she owed me a kiss. What might it have been? (Answer given later.) There is another statement I could have made which would have to be true and she would owe me a kiss. What statement would work? (Again, answer given later.) Now interestingly, there is a statement such that there is no immediate way of knowing whether it is true or false, but in either case she would owe me a kiss. Can you figure out what statement could work? (Answer given later.)

Anyhow, the important thing is that she owed me a kiss. That was a pretty sneaky way of stealing a kiss, wasn't it? Well, what happened next was even more interesting. Instead of collecting the kiss, I suggested we play for double or nothing. She, being a good sport, agreed. And so with another logic trick, she now owed me two kisses, then four, then eight, and things kept doubling and escalating and doubling and escalating, and before I knew it I was married! I've been married to Blanche, the charming lady musician, for forty-eight happy years. Unfortunately she passed away in 2006 at the remarkable age of one hundred! Even more remarkable, she looked no older that her late seventies!

Now for the answers to the three problems I raised above: The statement I made to Blanche was: "You will give me neither your autograph nor a kiss." If the statement was true, she would have to give me her autograph as agreed, but doing so would make it false that she gives me neither her autograph nor a kiss, which is a contradiction, and as the statement can't be true, it must be false. Since it is false that she gave me neither, then she must have given me either (false neither is the same as true either) and thus must give me either her autograph or a kiss, but the rule was that she can't give me her autograph for a false statement, and so she owed me a kiss.

What true statement would ensure that she give me a kiss? One statement that works is: You will either not give me your autograph or you will give me a kiss. The statement asserts that at least one of the following two alternatives holds:

(1) She will not give me her autograph.
(2) She will give me a kiss.

If the statement were false, then neither (1) nor (2) would hold, and then since (1) doesn't hold, its opposite holds, which means that she *does* give me her autograph, contrary to the agreement that she doesn't give me her autograph for a false statement. Thus the statement can't be false; it must be true. Since it is true, she must give me her autograph as agreed, but this falsifies (1). Thus (1) is false, but either (1) or (2) is true, hence it must be (2) that is true. And so she owes me a kiss (as well as her autograph).

Now for a statement which could be either true or false, but in either case, she owes me a kiss. A statement that works is: you will either give me both your autograph and a kiss, or give me neither one. The statement asserts that one of the following alternatives holds:

(1) She will give me both.
(2) She will give me neither.

Suppose the statement is true. Then one of the alternatives does hold. But also, she must give me her autograph for a true statement, and so alternative (2) cannot hold. Therefore it is (1) that holds and so she must then give me a kiss (as well as her autograph).

On the other hand, suppose the statement is false. Then neither (1) nor (2) holds, which means that she must give me one but not the other—either a kiss or her autograph, but not both. Which one must she give me? It can't be her autograph, since she can't give me her autograph for a false statement, and so she must give me a kiss.

In summary, if the statement is true, she must give me both a kiss and her autograph. If the statement is false, she must give me a kiss but not her autograph. Whether the statement is true or false depends entirely on what she does. She must give me a kiss and she has the option of giving me or not giving me her autograph. If she gives me her autograph, that makes the statement true. If she doesn't, that makes the statement false.

Logic can be an amazing thing! The logic I used to win a kiss from Blanche belongs to a genre that Blanche's son Dr. Jack Kotik wisely dubbed "Coercive Logic." I think that's a perfect description! I will give more examples of coercive logic later on.

Here is another way of stealing a kiss which I have used many times. I say to a girl: "I'll bet you that I can kiss you without touching you." In one particular case in which I tried this, the girl was highly intelligent and analytic, and asked me for a precise definition of kissing and touching. After I gave them, she said: "You mean that you will kiss me in accordance with your exact definition of kissing and you will not touch me, according to your exact definition of touching?" I replied that I would. That she said: "That's obviously impossible, so of course I will take the bet!" I then said: "It doesn't have to be a bet for money; let it just be a bet of honor." She agreed. I then told her to close here eyes, which she did. I then gave her a kiss and said: "I lose!"

Let me tell you a cute incident in relation to this: Once at a convention of puzzle makers, in honor of Martin Gardner, I met a very intelligent and attractive lady named Kate Jones, and told her that I would bet her that I could kiss her without touching her. She replied: "Oh, I know—you will probably lose!" When I congratulated her on her cleverness, she told me that she had recently heard a somewhat similar story—namely, that a man came into a bar with a friend who ordered a martini. The man then put a tumbler over the martini and said: "I'll bet you a quarter that I can drink that martini without removing the tumbler." The friend, thinking this impossible, took the bet. The man, then, removed the tumbler, drank the martini, and gave his friend a quarter!

I then told her a comparable story: A programmer and an engineer were sitting side by side on an airplane. They had the following conversation:

> PROGRAMMER: Would you like to play a game?
> ENGINEER: No, I want to sleep.
> PROGRAMMER: It's an interesting game!
> ENGINEER: No, I want to sleep.
> PROGRAMMER: You ask me a question, and if I don't know the answer, I pay you five dollars. Then I ask you a question and if you don't know the answer, you pay me five dollars.
> ENGINEER: No, I want to sleep.
> PROGRAMMER: I'll tell you what. If you don't know the answer to my question, you pay me five dollars, but if I don't know the answer to your question, I'll pay you fifty dollars.
> ENGINEER: OK. I'll ask you a question: What goes up the hill with three legs and comes down with four?

The programmer took out his laptop computer and worked on it for about an hour but got nowhere. And so he gave the engineer fifty dollars. The engineer put the money in his pocket and said nothing. A bit miffed, the programmer said: "Well, what's the answer?" The engineer then handed him five dollars.

Suppose I make you the following offer: I hand you two ten dollar bills and tell you that I am to make a statement. If the statement is false, then you must promise to give me back one of the bills and keep the other, but if the statement is true, you are to keep both bills. This sounds like a good deal, doesn't it? Would you accept the offer? If you did, you certainly shouldn't, for reasons you will soon see. At one lecture I gave to a group of logic students, I asked if any one would accept the above offer. One student accepted. I then handed him two ten dollar bills and made a statement such that the only way he could keep to the agreement was to pay me a thousand dollars! (Another instance of coercive logic.) Can you figure out what such a statement could be? I'll tell you later.

I then told the poor student that I felt really sorry that I played such a nasty con trick on him, and to make amends, I would give him back his thousand dollars providing he would answer me a yes/no question truthfully. He agreed. I then asked a question such that the student, according to the agreement, owed me, not a thousand dollars, but a million dollars! What statement would work? (I'll tell you later.)

At that point I said that I was thoroughly ashamed of myself for playing such a foul con game, and so I would now give him a chance to win back his million dollars in such a way that it was not possible for me to con him again—namely I would give him back his million dollars and again ask him a yes/no question, but this time, he would not even have to answer truthfully—his answer could be either true or false, as he pleased. "Obviously," I said, "there is no possible way I could con you now. Right?" The students all agreed that, under those circumstances, there is no way I could now con him. And so he agreed. However, there is a way I could con him! I phrased the question so that he now owed me, not a million dollars, but a billion dollars! How is that possible? (Answer given shortly.)

I then said: "Today I am in a generous mood, and so I will give you a fifty percent chance of winning your billion dollars back, but for this I charge a nickel extra! Are you willing to pay a nickel for a fifty percent chance of winning back your billion dollars?" (This of course gets a good general laugh.) Of course he accepted, and so I said to the class: "I am writing on this piece of paper a description of an event which may or may not take place in this room in the next fifteen minutes." After writing it down, I folded the paper and handed it to another student with the remark: "So that I can't use any slight and change what I wrote, I am giving it to you to hold for me, but don't read it!"

I then turned to the student who owed me one billion dollars and said: "Your job is to correctly predict whether the event will or will not take place. Obviously your chance of being correct is fifty percent, right?" He agreed. I continued, "If you guess correctly I will give you back your billion dollars, but if you guess wrongly, you'll still owe me a billion dollars plus a nickel. And so, without telling me I want you to think in your mind whether the event will or will not take place." I then handed him a pen and a piece of paper and said, "If you believe the event will take place, write YES; if you believe the event won't take place, write NO, and then fold the paper. I will turn my back so I can't see what you write." I then turned my back, and after a bit, I turned around again and asked, "Have you made your decision?" He replied that he had. I then said, "Then you're lost!"

The fact is that what I wrote was such that regardless of whether the student would write YES or NO, he would lose! Can you guess what I wrote? (I'll tell you shortly.)

Epilogue—At this point the student owed me a billion dollars plus a nickel. I then did a magic trick that allowed him to not pay me the billion dollars. "But you still owe me a nickel," I said. I continued, "But I'll tell you what. I'll give you your nickel back as a gift and claim a tax deduction."

I made a video of this lecture. I make lots of videos these days. I made one video which was a documentary of the life of my dear departed wife Blanche. She was indeed a musician—an excellent pianist and teacher who for years headed a music school in Manhattan, where, not only piano, but many other instruments were taught. She had a staff of some quite famous people such as Leonard Rose for cello, Sam Barron for flute and some members of the Juilliard Quartet. My documentary includes events of her childhood, and one event in particular is of especial interest. She was born in Ghent, Belgium in 1905. During World War I, when the Germans invaded Belgium, several German officers occupied the house of Blanche's parents. One officer in particular sat by little Blanche for hours and hours while she was practicing the piano. At one point the officer had a furlough, and went back to Germany. When he returned, in Blanche's own words: "He, my so-called *enemy*, brought me many beautiful rare musical manuscripts which we could never have afforded."

As a friend who saw this incident in my documentary, said: "In music there are no enemies."

How true! The incident bears a similarity to the scene in the movie *The Pianist*, in which which the Nazi officer hears this Jewish pianist who is trying to escape the Nazis—he hears him playing and then hides him from the Nazis and takes care of him.

This is also reminiscent of an incident I heard in my childhood from my parents: A burglar broke into a house and heard a little girl playing the piano so beautifully that it brought the burglar to tears. Instead of robbing the house, as he originally intended, he gave the little girl a beautiful ring and departed.

I am a true romantic at heart (I can't help it; it's genetic!) and the following true story appeals to my romantic nature. The story is about the great opera singer Lauritz Melchior.

Sometime around 1925 when he was sitting in his back yard in his house in Germany, a strange thing happened. At the time there was a very popular singer known as the "Mary Pickford of Germany." At one point she made a movie in which she had to parachute out of a plane. Well, the wind blew her off course, and she landed in Melchior's back yard, practically in his arms! Three days later they were engaged.

I really love the idea of a wife dropping out of the sky!

Speaking of musicians, you probably know the story of two jazz musicians who met in New York City, and one asked the other: "How do you get to Carnegie Hall?" The other answered: "Practice, man, practice!"

About 50 years ago, there was a popular violinist named Rubinoff. The story goes that Jascha Heifetz was late for a concert he was to give in Carnegie Hall, and rushed into an elevator, carrying his violin. The elevator operator said: "You can't use this elevator; it's the servants' elevator!" Heifetz relied: "Please, I'm Jascha Heifetz!" The elevator operator replied: "I don't give a good God damn if you're Rubinoff; you can't use this elevator!"

There was also the story told about a bass and drum jazz combo who played at a supper club. One day the boss said to the drummer: "I don't know what is wrong, but people have been complaining about the music!" The drummer replied: "I'll have to speak to the bass man about this!" Well, he spoke to the bass player and said: "Let's go about this scientifically! First I go onto the floor and play the drums alone and you listen. Then you go out alone and play the bass, and I'll listen." They did that, and after the bass player played, the drummer said: "Aha, just as I thought! Too much bass!"

I'm fond of nasty stories that shock people! For example, there is the story that a man went backstage after Rubinstein gave a concert and asked him: "Mr. Rubinstein, can I please have three autographs?" Rubinstein replied: "Certainly, but why do you want three?" The man replied: "Oh, I know someone who will trade three Rubinsteins for one Horowitz!"

In a somewhat similar vein, here is a story I made up: A man decides to play a practical joke on a friend who has just given a concert. He goes backstage and says: "Look, I think you played well; I don't care what people say!"

The following story is true: A friend of mine who is a famous musicologist once attended a piano recital of a friend of his who played a Beethoven sonata very badly. My friend later went backstage and said to him: "*You* couldn't have played it better!" The pianist did not see the ambiguity and felt highly complimented.

The pianist Leopold Godowsky was quite a wit, if his jokes were, albeit, a bit on the mean side. He was once at a concert with Rachmaninoff and the pianist at one point lost his memory. After the concert was over, the two were walking home and Rachmaninoff said: "Wasn't it terrible the notes he forgot?" Godowsky replied: "Not as bad as the notes he remembered!"

Once a composer friend of Godowsky called him up at about 2:00 A.M. and said: "You must come over and hear my last composition." "Your very last?" "Yes." "Good!"

My favorite Godowsky story is the one in which he visited another composer friend, and when he arrived, the composer was composing merrily away with operatic scores all over the piano. Godowsky said: "Oh, I thought you composed from memory!"

Many people have remarked about the correlation between music and mathematics—both interests and aptitude. Well, the great nineteenth century mathematician Felix Klein (who was evidently completely unmusical) was once at a party where they were discussing this very subject, and pointing out similarities between the two. At one point Klein said: "Gentlemen, I don't understand. Mathematics is beautiful!"

For those who love classical music I will continue with musical anecdotes.

The famous violinist Isaac Stern formed a celebrated trio with the equally famous pianist Eugene Istomin and cellist Leonard Rose. Now, in a trio, the three are of equal importance. Well, once they played at the White House. After they played, the president (whose name I would rather not mention) said: "I wish to congratulate Mr. Stern and his two accompanists."

Perhaps my favorite musical story is about Johannes Brahms and the amateur string quartet: It seems that Brahms knew four string players, who were very bad musicians, but such nice people that Brahms liked to associate with them. The four decided to surprise Brahms one day and spent six months assiduously practicing Brahms' latest string quartet. Then, one evening at a party, the first violinist said: "Johannes, come upstairs; we have a surprise for you!" The five went upstairs and the four took out their instruments and started to play Brahms' latest quartet. Well Brahms couldn't take it any longer, smiled, and started leaving the room. The first violinist rushed over to Brahms and asked: "How did we do? Were the tempos good?" Brahms replied: "Your tempos were all good. I think I liked *yours* the best!"

There is a story told about a pianist and a cellist who were rehearsing together. At one point the cellist said: "You're playing too loudly; I can't hear myself." The pianist replied: "Lucky you!"

A true story concerns the pianists Josef and Rosina Lhévinne. At one point, when they were rehearsing a sonata for two pianos, Rosina Lhévinne said to Josef: "You are playing too soft; I cannot hear you!" Josef replied: "How can you hear me when you are playing too loud!"

The music critic Sigmund Spaeth had some cute musical definitions. His definition of a fugue is that musical form in which one voice after another comes in, and one listener after another goes out.

When I was a music student, I once told my teacher that I wanted to be a Bach specialist. He said, "A specialist? Do you know Sigmund Spaeth's definition of a specialist? A specialist is one who does everything else worse!"

An excellent definition!

Some idiot music critic wrote about a piano recital: "His intonation was excellent." (For those who don't know this, "good intonation" means nothing more nor less than playing in tune. "Intonation" is only applicable to string instruments, and perhaps a few wind instruments, but certainly not applicable to a piano. A pianist's intonation is bound to be perfect, unless the piano itself is out of tune.)

The composer Anton Bruckner was incredibly naive! He once attended an orchestral concert and liked the conductor so much, that he later went back stage and gave the conductor a tip! The conductor was both highly amused and touched, and he made a hole in the coin and wore it as a necklace for many years.

The conductor Thomas Beecham had a fabulous memory. Once he memorized an opera score, he would not have to ever look at it again. Years later, he would still completely know it from memory. He was also extremely absent-minded. Once he was about to conduct an opera, and with baton in hand, he turned to the concertmaster and asked: "By the way, what opera are we doing tonight?"

He also had an acidic sense of humor: Once one of his lady opera singers introduced him to her ten-year-old son. Beecham asked the kid whether he could sing. When the kid said no, Beecham said: "Oh, I see it runs in the family."

Speaking of singers, there was Nelly Melba (after whom Melba toast was named) who had a perfect voice, but her singing was quite uninspired. The music critic Ernest Newman described her voice as "uninterestingly perfect and perfectly uninteresting."

Then there was the singer Adelina Patti, who had a wonderful voice, and was a hard-nosed businesswoman. Her manager once told her that the salary she demanded was more than that of the president of the United States, to which she replied: "Then get the president to sing for you!"

I really love that response!

Going back a couple of centuries, there was one singer who had one of the greatest voices in history, but she was totally unmusical. The way that the violinist Paganini describes her lack of musicality was priceless! "I went to her concert and was quite bored. She has a good voice and a good technique, but she's lacking in musical philosophy."

By the way, many famous music critics of the time said about Paganini that of course his virtuosity was great, but that the greatest thing about him was the soulful way he played slow movements. Indeed, the great Frederik Wieck (the father of Clara Schumann) said about him: "When he played the *Adagio*, it moved me more than any *bel canto* singer I have ever heard!"

I love the story told about the great Russian singer Chaliapin, who sang a low C, and a member of the audience yelled "Bravo" an octave lower.

When the cellist Gregor Piatagorski was in his late teens, he was quite a prankster. He once attached a cord to the cello of a dignified player of the Warsaw orchestra, having previously arranged for a friend behind the stage to hoist the instrument into the air just as the bow was about to touch it! He lost a week's salary over this prank, but felt it was worth it!

As is well known, Bach had two wives and twenty children. Well, a woman once read a biography of Bach, and said to a friend: "Did you know that Bach had twenty wives, and at night he would go up into the attic and practice on an old spinster?"

I like the incident that in Bach's day, a certain church was looking for an organist. At one point the minister wrote to a fellow minister: "Since we can't get anyone better, I guess we'll have to settle on Bach!"

In Beethoven's day, a critic wrote about the young Beethoven: "He will never amount to anything as a composer."

On the other hand, when Mozart heard young Beethoven, he said: "Watch this man! He is going to give the world really something!"

Beethoven once got into an argument with a prince. At one point the prince drew himself up stiffly and said: "Do you realize you are talking to a prince?" Beethoven replied: "There are many princes, but there is only one Beethoven!"

Riddle: Why was it difficult for Mozart and Beethoven to find their teacher? (Answer given later.)

The grandfather of Felix Mendelssohn, Moses Mendelssohn, was a well known philosopher in his day. The father of Felix, a successful banker, was to say: "I used to be known as the son of my illustrious father. Now I am known as the father of my illustrious son!"

A spiteful music critic once wrote a scathing review of both the playing and the compositions of Franz Liszt. Many friends of Liszt urged him to write a response, which he did. He wrote: "I am glad you don't like my music, and I agree with you entirely! I am happy that I have been helpful in bringing your name before the public."

As is well known, Clara Schumann had great contempt for the music and playing of Liszt; she thought of him as little more than a show-off. Liszt knew this. Nevertheless, the two were great friends. Liszt did have a good sense of humor, as the following incident will reveal.

Once when Clara Schumann had to give a concert, Liszt escorted her to the concert hall. When they looked in, they saw a very rowdy audience. Clara said: "What I have prepared is much too good for this crowd!" Upon which Liszt replied: "Then why don't you play some bad stuff by Liszt?"

The nineteenth century pianist Anton Rubinstein looked very much like Beethoven, and it was even rumored that he was Beethoven's illegitimate son (which was false, of course). He had a wonderful technique, but was extremely careless and played wrong notes all over the place! Nonetheless, his playing had so much fire and imagination that people loved it.

The great music teacher Theodore Leschetitsky, after hearing a concert of Rubinstein, said to him: "You must have a fabulous technique to be able to mess up the last movement that way!"

In more modern times, another great pianist who played many wrong notes was Edwin Fischer. I heard the story that the pianist Clara Haskil was once on a train with a fellow pianist and the two were remarking how many wrong notes Fischer played, without realizing that Fischer was in the same compartment. When the train stopped, Fischer said to the two ladies: "Could you please help me get my suitcase down from above? It is very heavy, you see, being full of all my wrong notes."

Edwin Fischer is one of my two favorite pianists, the other being Artur Schnabel. And now I will tell you many incidents about Schnabel, whom I had the good fortune to know.

Before I ever met Schnabel, once at my house, my friend the composer Leon Kirchner and I were listening to a magnificent Schnabel recording of the Schubert posthumous A major sonata. After having heard it, I jokingly suggested to Leon that we phone Schnabel to congratulate him for this performance. To my amazement, Leon picked up the phone and called him, while I listened to the conversation on another phone. Of course I was nervous about taking the time of the great pianist Artur Schnabel, but to our amazement, after congratulating Schnabel, he said "Yah, now this sonata is still a classic rather than a romantic sonata, and..." and he kept us on the phone for about an hour tracing the whole development of the sonata form!

To digress for a moment, I must tell you of a very funny incident. The Schubert sonata to which I just referred is one of my favorite pieces and I have played it in concerts. I once met a group of people and they asked me what I played. Being in a prankster-like mood at the time, I mischievously told them that one of my favorite pieces was the Schubert A major sonata—the posthumous one, you know, the one he wrote after he was dead. (Mind you, I said the one he *wrote* after he was dead, not the one *published* after he was dead.) One lady drew herself up stiffly and said: "You needn't tell us what the word means; we're college graduates."

I once played some of this sonata to Schnabel, and soon after, we got into a philosophical conversation in which at one point he said: "I am a realist, and because I am a realist I can sit back and be an idealist, because ideals are the reality!"

I read somewhere: "I am just a simple musician," said Schnabel, with the air of an emperor.

Schnabel really had quite a wit. Once in London, he came out from a dentist and told a friend: "I paid fifty shillings for these shifty fillings!" This was particularly remarkable for a man whose native language was not English!

Answer to the last riddle: The reason that Mozart and Beethoven couldn't find their teacher was that he was Haydn!

I was delighted to recently find out that the Bolivian government honored its violinist Jaime Laredo by issuing a postage stamp on which was written *La Ra Do*.

After one concert, a lady came back stage and asked Schnabel: "Could I ask you for your autograph?" Schnabel replied: "You already did, so you could." Actually, Schnabel was wrong. She did not ask him for his autograph; she asked him whether she could ask him for one.

Schnabel gave several lectures at the University of Chicago, and I had the good fortune of hearing them. At one of them, someone asked him what he thought of his latest review. He replied: "I don't read my reviews—at least those in America, because when they make a suggestion, I don't know what to do about it. Now in Germany it was different. I recall that in one concert I gave in Hamburg, the reviewer said that I played the first movement of the Brahms sonata too fast. I thought about the matter and realized that the man was right, and I knew what to do about it, and so the next time I publicly played it, I played it a little slower; but when the American reviewers say things like 'the trouble with Schnabel is that he doesn't put enough moonshine in his playing,' I simply don't know what to do about it."

I have told the above incident to several philosophers and remarked that what Schnabel said had somewhat the air of logical positivism—the doctrine that any statement which in principle could not be either verified or falsified was simply meaningless. According to the logical positivists, about ninety percent of all past philosophical writings are meaningless!

I must tell you an amusing incident in connection with this: In a certain inn not far from where I live, there is a very good library of philosophical books in the dining room. I once asked the lady proprietor how come these books were there. She replied: "They belonged to my ex-husband who is a philosopher—in fact a logical positivist. In fact it was logical positivism that broke up our marriage!" Quite amazed, I asked: "How could that be?" She replied: "Because just about whatever I would say, he would say is meaningless!"

At another lecture Schnabel said: "Can you imagine, Stravinsky actually said that music, to be great, must be cold and unemotional! And last Sunday I was having breakfast with Arnold Schoenberg, and I told him, 'can you imagine that Stravinsky actually published a statement which said that music to be great must be cold and unemotional?' And Schoenberg got furious, and said: 'I said that first!' "

Speaking of Stravinsky, he was once aboard ship with Toscanini and said to him: "I think that Beethoven was just a big bluff!" Toscanini turned his back on him and never spoke to him again.

A certain pianist whom I knew very well was once playing the Schubert Trout Quintet with some chamber players at a private concert in a house. The last movement starts with a theme played by the strings, and then the piano repeats the theme. I noticed that when my friend repeated the theme, there was a subtle, almost imperceptible smile on his face. I doubt that anyone else in the audience noticed it. After the concert was over and I was alone with my friend, I asked him why he smiled when he played that theme. He smiled broadly at me and said: "It's just that I played the theme so much better than they."

The pianist I just told you about was one of my former teachers. Two other pianists, both quite famous, were quite helpful to me before I gave a concert at Rockefeller University (in which I played the same Schubert A major sonata). One was Richard Goode, and the other, Alicia de Larrocha. To my great delight, I somewhere read that when Richard Goode was asked whether he ate after a concert, he replied: "If I played poorly, I console myself with a good meal. If I played well, I reward myself with a good meal."

I read with equal delight that when Alicia de Larrocha was a little girl, once during a lesson, her teacher interrupted her playing and said: "No, it should be played this way!" She said, "I understand," and sat down and played it exactly the same way as she did before. The teacher said: "No, no, it should be played this way." She replied: "I understand," and again played it exactly the same way again as before. After the third try, the teacher gave up.

It does my heart good to see that she had a mind of her own!

I was once at a party at which there was a concert pianist, whose name I will not mention. I asked him if he would play something for us, and he replied: "I play only at concerts. I don't like parlor pianists!"

By contrast, Alicia de Larrocha was kind enough to visit my wife and me for the purpose of letting me play for her the program I was soon to give at Rockefeller University. After having heard me play and making some useful suggestions, she sat down and played for us all evening long, and seemed most happy doing so! I will never forget how beautifully she played the Schumann Humoresque.

I was terribly sad to hear that she recently passed away (Sept. 2009). She was such a wonderful person!

The pianist Rudolph Serkin once said: "Many a pianist has been destroyed by not accurately following the score." With all respect to Mr. Serkin, I simply cannot agree. Indeed, it seems that composers are often far more tolerant of deviations from the score than are performers, and especially critics. As an example, a student of Beethoven once played one of his sonatas to him, and after she was finished, Beethoven said to her: "That is not at all what I had in mind, but I like your interpretation at least as well."

As another example, Chopin was once playing his Barcarolle to a group of friends and at a certain part he played pianissimo, instead of fortissimo, as was written. When one friend asked him why he did that, he said: "It can be played equally well dramatically loud or dramatically soft. I am better at playing soft, and so I play it dramatically soft."

Alicia de Larrocha was born with absolute pitch, but gradually lost it as she grew older. I had the same problem. My sense of absolute pitch was excellent during my childhood, but has become gradually distorted through the years. Now I hear everything three steps too high—I hear C as E flat. In connection with this, I must tell you a very amusing incident:

My friend the computer scientist Marvin Minsky and I were once driving in a car—he was driving and I was in the front seat. In the back seat were two gentlemen from Bell Telephone Labs. The conversation turned to absolute pitch. At one point Marvin said to the two in the back: "You know Ray here has absolute pitch." One of the two in the back asked me: "How accurate is your sense of absolute pitch?" For some reason, I didn't hear the question and so he said a little louder: "How accurate is your sense of absolute pitch?" at which Marvin turned around to the two and said: "I forgot to tell you, he's also deaf!"

Marvin really has quite a sense of humor. I published my first math paper when I was 35 years old. One friend of Marvin's who had read my paper said to him: "That was quite a good paper that Ray wrote!" at which Marvin replied: "Oh yes, at the age of thirty-five, Ray decided to become a child prodigy."

Can a person who is born deaf have absolute pitch? My guess is that he can, even though it might never be possible to verify it. One interesting case I know is about a policeman who was totally unmusical, never had any musical training, didn't know the name of any notes, nevertheless had absolute pitch. How was it known? Well, various police stations had their own radio signals, each one a different pitch. This policeman was the only one of his station who upon hearing a police signal on the radio, could tell which station it was.

The stony faced composer and pianist Sergei Rachmaninov was known for his apparent severity and many thought he had no sense of humor. But he certainly did have one—a wry one indeed! Once he and the violinist Fritz Kreisler were playing a violin and piano sonata in Carnegie Hall. At one point Kreisler lost his memory and went over to Rachmaninov and asked: "Where are we?" Rachmaninov replied: "Carnegie Hall."

Another very severe musician was the conductor George Szell. People would say about him that his bite was worse than his bark. He also had a wry sense of humor. Once the pianist Glenn Gould was about to rehearse a piano concerto with Szell and the orchestra, and before playing, fussed a long time with the piano chair—kept raising and lowering it. Szell waited patiently and at one point said: "Perhaps, Mr. Gould, if we sliced off just a quarter of an inch of your derrière, it would be just right!"

Another time Schnabel was rehearsing a Beethoven piano concerto with Szell and kept giving direction to the orchestra. At one point Szell said to Schnabel: "Artur, you forget I am here!" Schnabel replied: "Yes, you are here and I am here, and where is Beethoven?"

There is the story told about some musicians from the Boston Philharmonic who were out together rowing. One fell overboard and shouted: "I can't swim!" Another shouted back: "Fake it!"

I particularly like the story of a violinist of an orchestra who always had a pained expression on his face whenever he played. One day he was alone with the conductor and they had the following conversation:

> CONDUCTOR: Jake, what is bothering you? Don't you like the way I conduct?
> JAKE: That's alright.
> CONDUCTOR: Maybe you don't like the programs I choose?
> JAKE: They are OK.
> CONDUCTOR: Is it that you don't like the other players?
> JAKE: They are OK.
> CONDUCTOR: Am I not paying you enough salary?
> JAKE: You're paying me enough.
> CONDUCTOR: Then what is it that's bothering you?
> JAKE: I don't like music!

Now for a musical quiz: what composers have feline tendencies? Well, here are some:

> Kat-chuterian
> Milhaud (pronounced mee-oh)
> De Pussy
> Henry Purr-cel
> Modeste Meowski
> Claws Monteverdi
> Benjamin Kitten

Now name some composers with canine tendencies. The obvious one is Johann Sebastian Bark. There is also one who is not always canine, but frequently so:

> Jacques Often-bark

Also, there is one whose voice is sweet:

> Arnold Schön-bark

Then of course there is Ludvig van Bark-hoven. And one who is particularly artistic:

> Mutts-art, who also was a member of a wolf gang.

Then there is a pianist who is most canine when he is home:

> Wilhelm Bark-house

There is also a famous conductor who has some elephant-like characteristics:

> Arturo Tusk-anini

Then there are some watery musicians:

> Sir Thomas Beacham

Amy Beach
Edward Mac Towel
Francis Cool Lake
Sea Belina
Frederick Show Pond
And, best of all, Franz Joseph Hydrant

What composers gamble a lot?

Domenico Scarlottery
Ludwig van Bet-Often*

What composer has gangster-like tendencies?

Buxte Hoodlum

What composer belongs in a delicatessen?

Albaloney

Which composer is very fragile?

Benjamin Brittle

What musical form is very frightening?

The Scare-tzo

What composer always interfered in other people's affairs?

Felix Meddlesome

What composers have bovine tendencies?

Tchai-Cow-Ski
Rimski Cow-Sekoff

What composer is metallic?

---

*by Sylvie Degiez.

Johannes Bronze

What musicians are overheated?

Leonard Burn-stein
Van Cli-burn
Ludwig van Beet-oven

What composers are prone to colds?

Tchi-cough-ski
Pro-cough-ief
Rimski Korsa-cough

What composer obviously has a cold?

Rimski Of-course-a-cough

What musicians are wealthy?

Serge Rach-money-nov
Glenn Gold

What composer has enough money?

Rach-money-enough

Now for a literary quiz: What famous playwright resembles a weapon brandished back and forth?

William Shake-spear

What writer stings?

Nathaniel Haw-thorne

What author sharpens sticks?

Walt Whittle-man

What author stammers?

Ernest Hem-and-Haw

What playwright resembles a banana?

Eugene O'Peel

Now here are two little problems I fell for—I gave the wrong answer to both:

Suppose you have a boat with a metal ladder coming over the side. The ladder has six rungs spaced one foot apart. At low tide, the water hits the second rung from the bottom. Then the water rose two feet. Which rung did it then hit?

When given this problem, I responded: "The obvious answer is the fourth rung from the bottom, but that is too obvious to be right, I really cannot see what is wrong with my reasoning!"

What was wrong was that I totally forgot that the boat, being a floating object, rises with the water! Hence the water still hits the second rung from the bottom.

The second one I fell for is this: In a certain small town, thirteen percent of the inhabitants have unlisted phone numbers and no one has more than one phone. One day a statistician came into the town and took thirteen hundred numbers completely at random from the phone book. Roughly how many of them would you say have unlisted phones?

When I heard this, I took 13% of 1,300 and gave the answer 169. Of course I was wrong. The correct answer is zero, since the numbers were taken from the phone book!

I once gave this problem to a colleague, who immediately answered: "Obviously zero, since the names were taken from the phone book. What do you think I am, stupid?"

I really don't think it is stupidity that prevented me—as well as many highly intelligent people to whom I have given this problem—from getting it correct. I don't quite know what the faculty is that enables some people to get it and others not.

Here is another one I fell for: Someone once asked me how I was in grammar. I said I thought I was O.K. She then asked me: "If you have an egg, what is correct to say—the yolk ARE white, or the yolk IS white? I replied: Obviously, the yolk IS white." She said: "Wrong! The yolk is yellow!"

She then fooled me with another: Suppose you have a roof in which the sides are uniquely pitched. The west side is almost horizontal and the east side is almost vertical. Now suppose a rooster lays an egg on the peak of the roof. In which direction will it fall? I replied that I had no way of knowing. She replied: "Roosters don't lay eggs!"

Now, here are a few more that some of you might fall for. Answers follow.

1. In the state of Florida, is it legal for a man to marry his widow's sister?
2. If an airplane crashes on the border between Canada and the United States, in which of the countries should the survivors be buried?
3. Arthur and his father were out driving. The car got into a bad accident and the father was killed. Arthur was badly injured and was rushed to a hospital. The

chief surgeon took a look at him and said: "I can't operate on him; he's my son." How do you explain this?

Answers:

1. Many people to whom I have asked this, reply: "I don't see why not." The true answer is NO, because if a man has a widow, he is dead! (People often confuse "widow's sister" with "dead wife's sister.")
2. One doesn't bury survivors.
3. The surgeon was Arthur's mother.

I once attended a conference consisting of magicians and puzzle makers, and I heard one item which I thought was quite neat: Imagine you are living in a country that is very insecure in that if you mail a package, the contents are bound to be stolen unless you put a padlock on the box. Now, a man wants to mail his girlfriend an engagement ring. How can he do so in a secure manner? Of course he can put a padlock on the box, but she can't open it, since she doesn't have the key. Yet it can be done, if the box is sent back and forth. How? (Hint: it is possible to have more than one padlock on the box.) (Answers given later.)

Here is a very cute little puzzle to give to children. A certain man lived in the country. One day he had to go to the city for a couple of weeks on business, and before he left, he told his butler to forward any mail that came into his mailbox. The butler agreed to do so.

The man went away, but after a few days, he did not receive any mail, and so he phoned his butler and asked: "Didn't any mail come into my mail box?" The butler replied: "Yes, but I can't open the box because you took the key with you." The man replied: "Oh, no problem, I'll mail you the key in the morning."

Pretty stupid, huh? See if the kid gets it.

Answer to the Padlock problem: The man mails the ring with his padlock on the box. When the girl receives it, she puts her padlock on as well and mails the box back to the man with the two padlocks on it. When he receives it, he removes his padlock and then mails it back with just her padlock on it. Then, of course, she removes her padlock and opens the box.

I understand that this elegant problem comes from a Russian coding problem.

Here is a little problem that I believe quite instructive, particularly for children: Suppose you and I have the same amount of money. How much of my money must I give you so that you have ten dollars more than me?

Curiously enough, many even highly educated people give the wrong answer "Ten dollars." I then explain: "Well, suppose we have $100. If I give you ten you will then have $110 and I will have $90, hence you will have $20 more than me, not ten more than me!" Then the person usually gets it right and says "Five dollars."

One adult to whom I once gave this finally realized that five is the correct answer, but asked me how it works that way. I then gave the following explanation: Suppose we both have the same amount of money and an angel appears and takes five dollars out of my pocket and holds it in the air. At that point you have five dollars more than me. Then the angel puts those five dollars into your pocket, and you then have ten dollars more than me. This explanation satisfied him.

Speaking of explanations, when I was about ten years old, I asked my twenty-year-old brother Emile how the radio works. He gave me the following explanation:

Imagine a very long dog whose head is in New York and whose tail is in Chicago. You pinch the tail in Chicago and the head in New York barks. Well, you don't have the dog; instead you have the radio.

Emile really had quite a wit! Once I heard the grammar school joke: Where was Moses when the lights went out? Answer: he was in the dark. Well, I told this corny joke to my brother, who responded: "He couldn't have been in the dark because he was an Israelite."

How are you on biblical history? Do you know how many animals Moses took on the Ark? Now, don't just answer "Two of each kind," I want to know the *exact* number of animals all told. (Answer given later.)

There is the story told that God once visited Julius Caesar and said: "I have a commandment for you." Julius Caesar asked: "What is it, Lord?" God replied: "Thou shalt not kill!" Caesar replied: "Well, in theory, I'm with you, but politically, it's not very practical." The Lord was sad, and went to Marc Antony and said: "I have a commandment for you. Thou shalt not commit adultery!" Antony responded, "Now Lord, you know me well enough to know that I could never obey a commandment like that." The Lord was even more sad, and went back further in time to Moses and said: "I have a commandment for you." Moses said: "How much will it cost?" God replied: "Nothing. My commandments are free." Moses replied: "OK, I'll take ten."

Does God answer prayers? Well, a man once told his friend: "God doesn't answer prayers! I once prayed for a million dollars, and I never got it!" His friend replied: "God did answer your prayer. The answer was NO!"

Answer to biblical history question: Moses didn't have an ark.

I once had an idea for a cartoon: You see a picture of clouds in which there is a building named JUDGMENT CENTER and behind a desk is an angel with wings. In front of the desk stands and elderly couple, and the angel says: "I can't tell you yet whether you have salvation; our computers are down."

Some time ago I read an allegedly true story that I found quite sad: A minister who was about to be hanged was asked if he had any last words. He said: "I have nothing to say at this time."

On a more humorous note, a minister and his friend were playing golf. At one point the friend missed a hole, and said: "Missed the son of a bitch!" The minister said: "You shouldn't talk like that!" They then came to another hold, and again the friend said: "Missed the son of a bitch!" The minister said: "If you talk like that, Heaven will strike you dead!" The friend laughed at this, and when they came to another hole, the friend again missed, and again said: "Missed the son of a bitch!" Just then a bolt of lightning came down and struck a few feet from where he was standing, and a voice from Heaven said: "Missed the son of a bitch!"

Another humorous one: a girl in Rome had a date with her boyfriend in the evening, and in the morning she knelt by a statue of the Virgin Mary and prayed: "You who have conceived without sinning, please help me to sin without conceiving!"

A girl once came to a convent and told the Mother Superior that she wanted to become a nun. The Mother Superior asked her whether she had considered this for a long time. The girl answered: "No, last night I suddenly decided that want to become a nun." The Mother Superior said: "Now, this is a very serious matter, and we don't like people to consider this lightly. I suggest you go home and think about the matter very carefully for a year, and if at the end of that time you still want to join us, we will be happy to have you." The girl thanked her and went away. A few months later, the Mother Superior met the girl on the street all dressed in furs and jewels. She asked her: "Have you decided to become a nun?" The girl replied: "No. Instead I decided to become a prostitute." The Mother Superior angrily and excitedly said: "What did you say!?" The girl said: "I said that instead of becoming a nun, I decided to become a prostitute." With a sigh of relief, the Mother Superior said: "Oh, I thought you said Protestant!"

The following incident is true: A certain quite arrogant Catholic lady of prominence once visited the Pope and got into an argument with him. At one point the Pope said: "You forget, my dear, that I am also Catholic!"

To continue with theological jokes, a man fell off a cliff and there was a thousand foot drop. Fortunately he was able to grab onto a branch of a tree which was only a few feet down from the top. He hung there with both hands and had no idea how he could get back to the top. He finally raised his head and shouted, "Is there anyone up there who can help me?" A voice from Heaven came: "Do you believe in God?" The main desperately cried: "Yes!" Then the voice said: "Do you believe in the bible?" "Yes! Yes!" "Then just let go, and you will float down as gently as a bird." Well, the man still hung there, not knowing what to do, and then raised his head and cried: "Is there anyone else up there who can help me?"

There is the story of the world's ugliest man who was short, misshapen, and had no friends, and no girls liked him. One day he decided to do something about it! He went to an amazingly good plastic surgeon who stretched him out a couple of inches, moved over his nose and ears a bit, and did many other things which resulted in the man coming out tall and handsome! He now had plenty of friends, lots of girls who adored him, and one day he was out riding in a Cadillac with one girl in the front seat, three girls in the back seat, and suddenly a bolt of lightning struck him dead!

He went to heaven and said to God: "The first time I had an opportunity to enjoy life, you struck me dead. Why, Lord, why?" God replied in an agitated voice: "Jake, I didn't recognize you!"

The following story has always struck me as philosophically profound: a very holy man was in the ground floor of his house when a flood came. A boat then came by and the navigator said: "Come on, we can save you." The holy man replied: "I don't need saving. I am a man of God, and I know that He will take care of me. I will move up to the second floor." The boat then went away. A couple of days later, the flood rose to the level of the second floor, and another boat came by, and the navigator said: "For Heaven's sakes, come on, we can save you!" The holy man again told him that he trusted in God, and that he would move up to the third floor. The boat then went away. A few days later, flood reached the third floor, and a third boat came by, and again the holy man said that he needn't take it, since he was a man of God. The boat sadly went away, and the water reached the ceiling and the holy man drowned. He went to Heaven and said to God: "You saw I was in trouble—why didn't you try and help me?" God replied: "What do you mean I didn't try to help you? I sent three boats!"

What I find so enlightening in that story is that contrary to the common belief that God helps only through a miraculous manner, a God can help in a purely naturalistic fashion.

Another thing that I have wondered about is why people take it for granted that God favors those who believe in Him? Perhaps God is a scientific God who prefers beliefs based on evidence to those based on faith.

One of the most beautiful characterizations of God that I have ever come across occurs towards the end of Herman Hesse's novel *Knulp*. Knulp is freezing to death in the snow and is telling God how he regrets his life of vagabondage. Then God says: "Don't you see that whatever happened is good and right and that nothing should have been any different?... Look, I wanted you the way you are and no different. You were a wanderer in my name and wherever you went you brought the settled folk a little homesickness for freedom. In my name you did silly things and people scoffed at you. I myself was scoffed at in you and loved in you. You are my child and my brother and a part of me. There is nothing you have enjoyed and suffered that I have not enjoyed and suffered with you."

I have read that passage to several people, all of whom were deeply moved.

Now I come to something that really puzzles me: There is so much talk these days about the so-called conflict between evolution and intelligent design. Actually, this is a completely false dichotomy! There is no conflict between evolution and intelligent design—many people who believe in evolution believe that evolution itself was intelligently designed! I myself believe that. The real conflict is between evolution and *creationism,* which holds that humans did *not* evolve from lower life forms. But it is a serious mistake to identify intelligent design with creationism!

Those of you who wish to refute evolution can take one of two approaches: one is to assert that the scientific evidence for it is inadequate (as one fundamentalist once told me). Another is to admit that the scientific evidence supports evolution, but nevertheless evolutionism cannot be true because it goes against Scripture, and when there is a conflict between Science and Scripture, it is Scripture that should be trusted. As one religious tract has it: "Would you trust the word of scientists above the word of God?"

To those of you who are Christian fundamentalists and believe literally in a Heaven and Hell, I would like to ask the following question. Suppose that when you get to Heaven, to your surprise God says to you and the other saved ones: "I know that there has been much controversy as to the moral justification of Hell. Now, I have my own ideas on the matter, but I want you people to be happy, and so I want you to vote on the question as to whether I should abolish Hell or not. If more than fifty percent of you vote for the abolition of Hell, I will abolish it, otherwise I will retain it."

My question now is: how would you vote? I have asked this of many people, and the majority have said that they would vote for the abolition of Hell. Of those who voted for the retention, some have said: "After all, those were bad people!" Others have said: "Heaven and Hell are two sides of the same coin. You can't have one without the other."

The most clever answer I ever got was from a theology student at Notre Dame University, who referred to himself as an "orthodox Catholic." He said: "I would vote for the abolition, but that may well be an imperfection on my part."

One person to whom I posed the question said that the whole situation is unrealistic, since God would never form a judgment based on people's opinions. However, I believe there is scriptural support against what he said, because in the Old Testament, when God decided to kill all the Israelites, Moses did change his mind!

Another question I like to pose to believers is about that which I call "Collective Salvation," which is that on the day of judgment, God takes the average of all the past deeds of the entire human race, and we either all rise or all fall together! How would you like that scheme? One person to whom I asked this said: "Ooh! That's a dangerous idea!" To my great amusement, another one said: "I don't like that idea at all! I think my chances would be much less!"

Prior to all this, one should of course ask the more fundamental question of whether God exists. One friend of mine gave me the following proof that God does *not* exist. He said: "I now have absolutely irrefutable evidence that God does not exist. How do I know that God doesn't exist? Why, God himself told me so, and surely He should know!"

Of course my friend was jesting, but interestingly enough, a freshman student at Indiana University, where I once taught, quite seriously gave the following proof that God *does* exist: "God must exist, because He wouldn't be so mean as to make me believe that He exists if He really doesn't."

While I was at Indiana University, one of the courses I taught was a freshman logic course for liberal arts students. In one particular class, I wanted to give those students who were poor in logic a chance to raise their grade, and so on my final exam, one of my questions was to write anything you want. Well, one student wrote a story that I found so remarkable, that had he been a failing student, I would have given him an A+ just on the basis of the story. Actually he was an A+ student anyhow, but I can assure you that had he been an F student, I would have given him an A+ anyhow. Here is the story he wrote:

Once upon a time, there were two tribes—the selfish tribe and the altruistic tribe. Those of the altruistic tribe never wanted to do things for themselves, they wanted to do only things for society. Those of the selfish tribe, not only didn't want to do things for society—they had a positive aversion to doing things for society. Well, one day these two tribes got into a war and all were killed except one member of each. These two then faced each other with drawn guns, and the altruistic one thought: "If I kill him, I'll be the whole of society, and anything I do for society I'll be doing for myself, and I don't want to be selfish!" At the same time, the selfish one thought: "If I kill him, I'll be the whole of society, and anything I do for myself I'll be doing for society, and I don't want to do anything for society!" As a result, neither one shot the other.

That was the story. I recall that when the student handed me the paper, he said: "Professor Smullyan, I think you will like this." I sure as hell did!

Now let me ask you a question: Is it possible for the words *yes* and *no* to be used synonymously? I can think of a case in which they can: "He's not a nice guy!" "Oh yes." "He's not a nice guy!" "Oh no!" In both cases the words *yes* and *no* are used affirmatively.

The famous mathematician Stanislaw Ulam thought of the following paradox, which is now know as the Ulam Paradox: When President Richard Nixon was appointed to office, on the first day he met his cabinet he said to them: "None of you are yes-men, are you?" And they all said, "NO!"

This resembles a cartoon a student once sent me of the master of his house saying to his very frightened-looking butler: "I hate yes-men, Jeeves, don't you?"

The same student once sent me the message inside a Chinese fortune cookie, which was: "Do not depart from the path that fate has chosen for you."

The philosopher Sidney Morgenbesser was once in England and attended a lecture on linguistics. At one point the lecturer said, "In many languages, a double negative makes a positive; but in no language does a double positive make a negative." At which Morgenbesser said, "Yeah, yeah!"

Sidney Morgenbesser was once with a group of other philosophers who were discussing Kant's ethical principal that *ought* implies *can* (meaning that is is unreasonable to expect a person to do something which is impossible for him to do). Morgenbesser, with his typical sense of humor, then said: "Jewish ethics is different. In Jewish ethics, *can* implies *don't*!" Then, he gave an explanation of Jewish logic and Jewish epistemology. Jewish logic is: if P, why not Q? And Jewish epistemology is: "Mother knows best!"

Which brings me to some of my favorite Jewish jokes.

A New Yorker was away from New York for a year. When he returned, he went to his favorite Jewish restaurant and to his amazement had a Chinese waiter who took his order in perfect Yiddish. Later the man asked the owner: "How come this Chinaman is learning Yiddish?" The owner replied in a soft voice: "Shhhh—he thinks he's learning English!"

An orthodox Jew was in China on the eve of an important Jewish holiday, and looked for a synagogue in which to pray. He finally found one, and when he entered he saw all these Chinamen praying. The leader came over to him and said: "It is very nice you come visit with us, but today not good day for visit since today is very holy day and we pray." The man replied: "I didn't come to visit; I came to pray with you." The leader with a puzzled expression asked: "Are you Jewish?" The man replied: "Yes." The leader said: "That's very funny, you don't look Jewish!"

Mrs. Goldberg wanted to visit an important Lama in Tibet. She got up the mountain with the aid of sherpas, and knocked on the door of the temple, which was opened by a man. She said that she wished to speak with the Lama. The man said, "Sorry, but the Lama doesn't speak to women." She said: "Tell him that I want to say only three words." He went away and came back a few minutes later and told her that under those circumstances, the Lama would see her. She then came into the room where he was and said: "Sheldon, come home!"

A Jewish mother gave two ties to her son for a birthday present. The next day the son came to the mother wearing one of the ties. The mother looked at him and said: "What's the matter, darling, you didn't like the other tie?"

Two lady friends met on the street and one of them asked: "What's new?" The other replied, in an agitated voice: "What's new? Why, my husband is in the hospital and the doctors don't know what is wrong with him. He's growing a tail!" The other replied: "Growing a tail? So what else is new?"

Abe and Isaak met in the street.

ABE: How's your health?
ISAAK: Fine.
ABE: How's Becky?
ISAAK: She's doing fine.
ABE: And Joseph?
ISAAK: Also good. How come you don't ask me how is business?
ABE: Alright, how is business?
ISAAK: Don't ask!

A wife said to her husband: "Please close the window. It's cold outside." He closed it and said: "So now it's warm outside?"

Two rabbis were sitting on a bench. One said: "Life is tragic!" The other said: "Of course life is tragic. One is lucky if he's never born!" The other said: "Of course one is lucky if he is never born, but how many are that lucky? Maybe one out of a million!"

During a service in a synagogue, the rabbi beat his breast and said: "I'm nothing! I'm nothing!" and fell to the floor. Then the temple singer followed suit and said: "I'm nothing! I'm nothing!" and fell to the floor next to the rabbi. Then the temple servant decided to get into the act, and said "I'm nuttin'! I'm nuttin'!" and fell to the floor. Then the rabbi nudged the singer and said: "Look who thinks he's nothing!"

An orthodox Jewish student asked his rabbi why he had to wear a yarmulke. The rabbi said: "Of course you have to wear a yarmulke!" The student replied: "I know you have to, but you haven't told me why. Why?" "Because it says so in the Book, that's why!" "But I have read the Book, and found nothing there which explains why." "Then go home and read it again." The student went home and came back some months later and said: "I read the Book most thoroughly, and nowhere could I find any reference to the yarmulke. Where does it say that one should wear a yarmulke?" "It is everywhere in the Book!" said the rabbi. "Just open to any page at random." The student opened a page and the rabbi said: "Read! What do you see?" "I see that it says that Moses then went forth into the desert." Rabbi: "And you think Moses would go into the desert without a yarmulke?"

The following Jewish story (perhaps my favorite) was told to me by one who is not Jewish.

A rabbi was being chauffeured to another city where he was to give a presentation at a temple. The chauffeur was acting very disgruntled. The rabbi asked him what was troubling him. The chauffeur replied: "Oh, you work so little and get so much admiration! I work so hard and get so little!" The rabbi replied: "If you feel that way, let's change roles." And so the rabbi moved to the front and drove and the chauffeur sat in the back.

When they reached the other city, the members of the congregation, who had never seen the rabbi before, took it for granted that the one in the back seat was the rabbi, and so they took him into the temple, and the real rabbi followed, in the role of the chauffeur. They then asked the chauffeur, who they thought was the rabbi, a difficult theological question. The chauffeur then pointed to the real rabbi and said: "Oh, that's so obvious, even my chauffeur could answer that!"

An Armenian once sold a horse to a Jew. An observer on the scene wanted to find out which one got the better of the deal. First he went to the Jew and asked him how much he paid. The Jew named a surprisingly low figure. The observer then went to the Armenian and said: "How come you sold such a magnificent animal for so little?" The Armenian replied: "He got no bargain, the horse is lame!" The observer went back to the Jew and said: "Aha! He got the better of you. The horse is lame!" The Jew replied: "The horse is not really lame. I noticed a nail sticking in his foot which makes him limp, but when I take the nail out, he will be fine!" The observer went back to the Armenian and said: "You may not know it, but that horse whom you thought was lame isn't really lame. There is a nail sticking in his foot which makes him *appear* to be lame!" The Armenian replied: "No, the horse really *is* lame. I just stuck the nail there to make the Jew *think* the nail was the cause of the limping, but he will find out that the horse really is lame!" The observer went back to the Jew and said: "He really fooled you! The horse really is lame. The Armenian put the nail there himself to make you think the horse wasn't really lame." The Jew replied: "I thought of that possibility, and that's why I paid him in counterfeit money."

An Irishman, Jim Murphy, went into a bar and ordered a beer. After drinking it, just as he was walking out, the bartender said: "Hey, wait a minute! You haven't paid me!" Jim said: "I certainly did!" The bartender said: "I don't recollect you paying me." Jim said: "Well, can I help it if your memory is short-lived? And besides, the customer is always right, isn't that so?" The bartender was thoroughly confused and said nothing. Jim walked out and met Sean, an Irish friend, and told him what happened, and that Sean could do the same thing. "Just tell him that the customer is always right." And so Sean went into the bar and pulled the same trick. The bartender was furious, but didn't know what to say. Sean walked out and met his Jewish friend Abe Goldstein and told him what happened, and that he could do the same thing. Abe went into the bar, ordered a beer and drank up. The bartender said: "You know what just happened? Two guys walked into this bar and ordered beers and didn't pay for it and both claimed that they did. Now, the next guy who tries a trick like that, I'm just going to twist his head around his neck, see?" Abe said: "Look, don't bother me with your troubles, just give me my change, I want to go!"

A man went into a store and asked for a pack of Camels. The clerk brought one and put it on the counter. The man said: "On second thought, I think I would prefer to have a pack of Chesterfields." The clerk put a pack of Chesterfields next to the pack of Camels. The man picked up the pack of Chesterfields and said: "I'll trade you the Camels for the Chesterfields," and started walking away. The clerk said: "Hey, you didn't pay me for the Chesterfields!" The man replied: "I gave you Camels in exchange for them!" The clerk said: "But you didn't pay me for the Camels!" The man replied: "I'm not taking the Camels; I'm taking the Chesterfields."

I love clever swindles! The most elegant one I know is known as the "Ten Dollar Bar Swindle." It was performed several times at about the end of the nineteenth century. It has to be done at a large bar in which there are two cash registers quite far apart.

A magician goes into the bar and entertains many of the customers as well as the bartender. At one point he announces that he is now about to perform his greatest trick. He asks to borrow a ten dollar bill from the cash register. The bartender obliges. The magician then asks him to put his initials on the bill so that the bill can be identified. The bartender does that. The magician then puts the bill into an envelope, in full view of the all the spectators. However, what the spectators don't know is that the envelope has a slit in the back. The magician gets the bill out of the slot and palms it and hands it to a confederate. None of the spectators know that he has a confederate. The magician says: "Now I will burn up the bill in this envelope." As he makes preparations for the burning, the confederate goes around to the other cash register, hands the ten dollar bill to the other bartender and orders a martini. The other bartender puts the bill in the register, takes out nine dollars which he gives to the confederate. Now the magician sets fire to the envelope and after it has completely burned out, says: "Now I will magically restore the bill and make it travel. You will find it in the other cash register over there!" Someone goes over to the other register, brings back the bill, and sure enough, the bartender's initials are on it! Everyone yells BRAVO!

Now let me tell you of a magic trick that I consider to be the ultimate in elegance! It was done by Sam Loyd (1841 – 1911), who was one of the greatest puzzle inventors the world has ever seen, and also a magician. He often did this trick aboard ship, and it fooled even magicians—no one could figure it out! He had his twelve-year-old son blindfolded and his back turned to the audience. One of the spectators had his own deck of cards, shuffled them himself, and showed them one by one to Sam Loyd and the boy would correctly name the card each time! How was this done? This was performed in the days before radio signals were invented, and as I told you, even magicians couldn't figure it out. I will tell you how it was done after the next three items.

Another trick that was done around the turn of the twentieth century: A magician walked into a small town where he was unknown. He was allowed to go in the afternoon to the theater in which he was to perform in the evening and he could make his preparations. Well, when evening came, the audience was full. All that could be seen on the stage was the magician seated behind a table upon which was a deck of cards in a cardboard case. A volunteer from the audience (who was not an assistant or confederate) came up to the stage, walked off with the pack of cards, took the cards out of the case, examined them and testified that they were all different, shuffled them up and had three members of the audience each take a card and hide it in his pocket or her purse. The volunteer then put the forty-nine remaining cards (there were no jokers) back in the case, which he closed, and then went up to the stage and placed the closed pack on the table. The magician then correctly named the three missing cards.

How was this done? I have presented this to many people as a twenty questions game, and I will give you some hints: One person wanted to find out at what point the magician knew which cards were missing. Well, he knew it only after the pack was put back on the table. By which sense did the magician know? He knew by the sense of sight. Even though the pack was closed, he saw something that enabled him to know the missing cards. Were the cards ordinary? No, they were not, even though no one looking at them, either fronts or backs, could see anything wrong. Was weight involved? No. And so how was it done? I'll tell you after the next two items.

It is amazing to what lengths a magician will go to prepare for a trick. In one such trick, a card was selected and returned to the deck. Then a basket of eggs was brought in and one was selected. The shell was broken open and inside the yolk of the egg was a little slip of paper in which was written the name of the selected card.

The hardest part of the trick was getting the slip of paper inside the yolk of the egg. How was that done? I'll tell you sometime after the next item.

Houdini was once sent to a country to tame a savage tribe. On the floor of his hut was a wooden trunk with a handle. The strongest members of the tribe were unable to lift the trunk from the floor, but when Houdini passed his hand over a little boy and said the magic words, the boy could lift it easily. How was this done?

Here are the answers to the four tricks described above.

1. In Sam Loyd's trick, the kid never said a word. Sam Loyd was a ventriloquist! I can't think of a more deceptive idea! A funny thing is that on one occasion, an elderly gentleman said to Sam Loyd: "You shouldn't tax the boy's mind that much, it's not good for him!"
2. In this trick, the magician had an x-ray machine hidden up in the ceiling. On the back of each card was a dab of metallic salt through which X-rays could not penetrate. The dabs were in different positions on the different cards. With the deck full, no X-rays could get through. But when three cards were missing, there were three holes which X-rays could penetrate. The X-rays went to a fluorescent screen hidden behind the magician's table, and by seeing the three spots of light, the magician could identify the three cards.
3. It was done by previously taking a live chicken and implanting the tiny slip of paper into the chicken's

ovary. The egg then grew around it. Good God, what an idea! I would imagine that a more humane method could be found.

4. At the bottom of the trunk there was a sheet of metal covered over by thin wood, and there was an electromagnet beneath the floor, which Houdini could control.

Now for some more jokes. A man working in a defense plant during wartime one day started walking out with a wheelbarrow full of hay. As he came to the exit gate, the guard said: "Halt! What are you hiding in the hay?" The man said: "Nothing." The guard said: "I will search!" He did so and found nothing hidden. He scratched his head and let the man pass. The same thing happened the second day, then the third, and this continued for six months. Then one day the guard could contain his curiosity no longer, and said to the man: "Look, I know you're up to something, but I can't figure out what! If you tell me what it is that you are stealing I promise not to report it." The man said: "Wheelbarrows."

Next, a man once walked into a bar with two carrots in his ears. The bartender was amazed, but said nothing. The same thing happened again for fourteen more days. On the sixteenth day, the man came in with celery stalks in his ears. The bartender could no longer control his curiosity and asked: "Why do you have celery stalks in your ears?" The man replied: "I couldn't find any more carrots."

Then there was the man walking on the street with bananas in his ears. One woman stopped him and asked: "Why do you have bananas in your ears?" The man replied: "Eh? Can't hear you; I've got bananas in my ears!"

Then there was a man in a restaurant who saw a man rubbing spinach in his face and beard. When asked why he was rubbing spinach in his face, he said: "Spinach? My God, I thought it was mashed potatoes!"

Then one day a reindeer walked into a bar and ordered a martini. He gave the bartender a ten dollar bill and the bartender gave him two dollars in change, and said: "Pardon me for staring at you, but one doesn't see many reindeer come into this bar and order martinis!" The reindeer replied: "And you're not going to see many more at these prices!"

A man once gave a friend a gift beautifully wrapped in a box. The friend opened the box, and in it was a live lobster. "Good," he said, "I'll take him home for dinner." The man replied: "No, no, he's already had dinner, just take him to a movie."

A theatrical agent once got a telephone call from someone who wanted a job. The agent asked him what he could do. The voice said: "I can talk!" The agent said: "What's so unusual about that?" The voice said: "You don't understand; I'm a horse!"

A man once brought a dog into the office of a theatrical agent. The dog took a violin out of its case and played. The agent shook his head and said: "A Heifetz he'll never be!"

Another man took his dog to a theatrical agent and proudly said: "Now look at what this dog can do!" He then said to the dog: "What is on top of a house?" The dog said: "Roof!" The man said: "What baseball player was famous?" The dog said: "Ruth!" The agent was not impressed. The two walked out and after a sad silence, the dog said: "Perhaps I should have said DiMaggio."

A man visited a friend, and to his amazement saw him playing chess with his dog. After expressing his amazement, the friend said: "Oh, he's really not all that good. I beat him two out of three."

Three tortoises were at a bar and ordered beer. One of them said: "This beer would taste good if we had some salt." One of the three said to another: "You being the junior member should go out and get us some salt." The junior one said: "No, if I get the salt, then while I will be gone, you'll drink my beer." Another said: "No, we promise not to drink your beer if you get the salt." The junior one said: "Alright, if you don't drink my beer, I'll get some salt," and hobbled away. A day passed, and he didn't come back. Two days passed and he didn't come back. Two years passed, and he didn't come back. One said to the other: "Do you think he'll ever come back?" The other said: "No, we might as well drink his beer." They started to do so, when the junior tortoise popped out from his hiding place and said: "Ah! Ah! If you touch my beer, I won't get the salt!"

A man once brought into a bar a mouse, a bird, and a tiny, tiny piano. The mouse sat down at the piano and played the *Hammerklavier* sonata. Then the bird sang the entire Schubert *Winterreise* with the mouse accompanying her. The bartender was utterly amazed and offered the man fifty thousand dollars to buy the act. The man agreed and the bartender took out his check book and was about to write a check, when the man said: "No, no, I can't do it! I can't do it! My conscience will bother me. The whole thing is a fake. A fake! That bird can't really sing! The mouse is a ventriloquist."

An American visiting England decided to partake of some typical British sport—fox hunting. And so he went on a fox hunt. After it was over, he asked the guide whether he comported himself in a truly British manner. The guide said: "On the whole you did pretty well, only when you see the fox, you should say: 'Tally Ho, the Fox!' not 'There goes the son of a bitch!' "

An American visiting France decided to partake of some typical French sport—rabbit hunting. And so he went out in the morning with a French guide and didn't see any rabbits until about twilight, when a rabbit popped out of a hole. The American raised his gun and was about to shoot, when the Frenchman said: "Do not shoot, Monsieur! That is Fifi, we never shoot Fifi!" And so no rabbits that day.

The next day the two went out again and in the late afternoon a rabbit appeared. The American raised his gun, this time a bit more slowly, when the Frenchman said: "No, do not shoot! That's Mimi; we never shoot Mimi!"

On the third day, when a rabbit appeared, the American did not even bother to raise his gun. The Frenchman said: "Shoot, Monsieur, shoot! That's Alfonse. We always shoot Alfonse!"

A Pole, a German and a Czech went out hunting. They did not come back at an expected time, and so a search party went out to look for them. They came across two enormous bears, a male and female, who looked as if they had just had enormous meals. They shot the female, and sure enough, the Pole and the German were found inside. At which one of the party said: "That means the Czech is in the male."

Two cows were standing in a field. One of them said to the other: "Aren't you worried about mad cow disease?" The other replied: "Why should I worry, I'm a helicopter!"

A lady went into a sweater shop and pointed to one she liked. The sales lady said: "It's not virgin wool." The lady said: "I don't care what the sheep has been doing; I like that sweater!"

In the country of Charlil, the inhabitants are reputed to be stupid. According to one story, the president of Charlil wanted to convince the citizens that athletes were not as stupid as was believed. And so he had thousands of people gathered at a stadium and had an athlete seated on the stage. He said to the audience: "To show you that athletes are not necessarily stupid, I will ask him a question." He turned to the athlete and asked: "How much is eight times eight?" The athlete thought very hard and finally said: "Sixty Four," at which the audience yelled: "Give him another chance!"

Do you know the definition of a Charlilian pianist? He is one who gives his concerts on a silent keyboard so that no one can hear his mistakes.

A Charlilian once returned a new book to the bookstore and complained that there were two things wrong with it. First, the pages were bound in reverse order, and second, the letters were printed upside down.

One Charlilian invented a remarkable copying machine that does double duty: as it makes a copy, it shreds the original.

An American, an Englishman and a Charlilian were about to be executed on the guillotine. First they asked the Englishman whether he wanted to die face up or face down. He bravely said "Face up." And so he lay on his back. The blade came down and got stuck half way down. This was a sign from God, and so they had to let him go. Then the American also decided to die face up, and this time the blade stopped only a couple of inches before his neck, and so they also had to let him go. Then the Charlilian also decided to die face up, and there he was looking up and suddenly said: "Oh, I see what's wrong with that guillotine!"

Two Charlilians were in a space ship. One of them went out for a space walk. When he came back, the door was locked; and so he knocked on it. The one inside said: "Who's there?"

One Charlilian wanted to commit suicide. He threw himself to the ground and missed!

Once a ventriloquist was entertaining at a bar and telling many anti-Charlil jokes. One customer got very upset and said: "Hey, don't you know that many of us here are Charlilian?" The ventriloquist began to apologize, upon which the Charlilian said: "I'm not talking to you! I'm talking to the little guy sitting on your lap."

Two Charlilian carpenters were working on a house. They had to panel a wall. One of them kept throwing about half the nails away. The other asked him why he was doing that. The first said: "Because the heads are on the wrong end." The other said: "You idiot! You should save them for the opposite wall!"

The last joke bears a resemblance to Texas Agi jokes that I heard when I was in Indiana. Texas Agies—students of the Texas Agricultural College—are reputed to be stupid. In one joke, a Texas Agi wood-chopper could chop down six trees a day. Then he saw an ad for a chain-saw which guaranteed that with it, one could chop down twelve trees a day. He bought it and came back a couple of days later and complained that with it he cut down only two trees a day. The salesman said: "Let's see what's wrong," and pulled the string to start the motor, at which the Agi said, "What's that noise?"

Another Texas Agi decided to raise a chicken farm, and so he bought a bunch of chickens and planted them, but nothing came up. He thought the trouble might be that he planted them heads up, so he bought another batch and planted them heads down, but still, nothing came up. He then wrote a letter to the Texas Agricultural College and explained very carefully what he had done and asked why there were no results. Two weeks later he got a letter from them saying: "Send soil sample."

My favorite Texas Agi story is about an Agi who read an ad in the papers: "You want a free trip to the Bahamas? Come to Pier 15 on March 3rd and we will take you there without charge." He went there on the right day and saw a long narrow boat with oars coming out of the sides. At the head of the pier was a Nubian holding a whip. He escorted the Agi into the boat, put him on a seat and chained him up. The Agi saw that the whole boat was filled with chained-up Agies. Then the Nubian started beating them all with his whip and they all began to row. The Agi said to his neighbor: "Do you think we have to tip him?" The neighbor replied: "We didn't last time."

Speaking of Texas, two Texans went into a car dealer and bought two gold-plated Cadillacs. One took out his check book and was about to write a check for both, when the other said: "No, no; it's on me. You paid for lunch."

In Indiana they make jokes about Kentucky, for example: they sent their lowest grade moron to Kentucky, thus raising the average I.Q. of both states.

Have any of you heard any Graf Bobby jokes? He was a European aristocrat reputed to be stupid and there are many jokes told about him. For example, on one occasion he was out walking with a walking stick with an ivory handle. He met his cousin who said: "What a beautiful walking stick!" Bobby said: "I don't like it; it's too long." The cousin said: "Then why don't you cut off a piece?" Bobby said: "What? I should cut off this beautiful handle?" "Of course not! Cut off a piece from the bottom." Bobby replied: "But that's not where it's too long!"

On another occasion, Bobby went to a doctor because of a rash on his face. The doctor suggested he put on some toilet water. Bobby came back a week later with bumps on his head. He told the doctor: "I used toilet water, but the seat kept falling down."

Bobby once bought a discarded railway car and used it as a house. One day his cousin came to visit him and saw him outside the car, sitting by a tree in the rain and smoking a cigar. The cousin asked him why he was smoking outside instead of inside the car. Bobby explained that there was a sign inside saying that it was not a smoking car.

On another occasion Bobby met his cousin on the street who told him: "Bobby, do you realize that you are wearing one black shoe and one brown shoe?" Bobby looked down at his feet and said: "Oh, I have another pair just like this at home."

Once Bobby visited his cousin and told him that he was expecting a baby and was worried that the baby might be Jewish. When asked why, he said: "I just read that one out of three who are born in our city are Jewish."

Once someone phoned Bobby at two o'clock in the morning and apologized for calling at such an hour. Bobby said: "Oh, it's perfectly alright. I had to get up and answer the phone anyway."

Once Bobby met his cousin after turning a corner around a cheese factory. The cousin said: "Oh, its you, Bobby." Bobby replied: "No, it's the cheese."

Bobby once visited New York City and at a party heard the following joke: A teacher said to a class: "Anyone who correctly answers my first question won't have to answer any other questions. How many hairs are there on a horse's tail?" One student raised his hand and said: "Two thousand three hundred and forty eight." The teacher asked: "How do you know?" The student replied: "That, teacher, is already your second question!"

Well, soon after, Bobby went to San Francisco and at a party said: "When I was in New York, I heard a joke. A teacher told his students that anyone who could correctly answer his first question wouldn't have to answer any more questions. He then asked how many hairs were on a horse's tail. A student then answered—just a minute now, let me make a phone call." He then made a long distance call to New York to the host of the former party and asked: "In that joke you told, how many hairs did the student say were on the horse's tail?"

Now for some definitions: What is the definition of a sadist? Answer: One who is kind to a masochist.

Next I would define an editor as someone whose function it is to ruin an author's manuscript.

Contrary to the usual definition of the word *oxymoron*, I would give the following two definitions:

1. A stupid ox
2. A moron educated at Oxford.

The German word *Leiermann* (pronounced *lyerman*) is usually translated as "organ grinder." I believe that is incorrect! To me, a *Leiermann* is a man who doesn't tell the truth.

How would you define the difference between an optimist and an incurable optimist? I would say that an optimist is one who says: "Everything is for the best. Mankind will survive." An incurable optimist is one who says: "Everything is for the best. Mankind will survive. And even if mankind doesn't survive, it's still for the best."

Do you know the difference between an optimist and a pessimist? The optimist says that this is the best of all possible worlds, and the pessimist sadly agrees!

Of course the most famous philosopher of pessimism was Schopenhauer. About him, William James said: "He reminds me of a barking dog who would rather see the world ten times worse than it is, than lose his chance of barking at it!" I asked a famous philosopher why it is that when I read the pessimistic philosophers, instead of feeling depressed, I feel elated. Why is that? He replied: "Of course, because you know it isn't true." I did indeed find Schopenhauer's defense of pessimism quite unconvincing, in fact quite childish. His two main theses are first that when one satisfies a desire, a new desire arises. Well of course, so what? Secondly, he points out how mendacious people are and how badly they generally act. This led me to think that there are really two types of pessimists, which I can appropriately call *contingent* pessimists and *essential* pessimists. The first claims that existence happens to be bad. The second claims that existence *has* to be bad—it is impossible for it not to be. Schopenhauer vacillates between the two. When he talks about how bad people are, he is a contingent pessimist. When he says that whenever we fulfill a desire, a new one comes one, then he is an essential pessimist. I found his defense of both quite inadequate. A far more striking and convincing one is Schopenhauer's successor von Hartmann, who, surprisingly enough, is not nearly as well known. Von Hartmann is what I would call an essential pessimist, with a surprising optimistic twist, that I will soon explain. Now, Schopenhauer raises the question of why we shouldn't all commit suicide, considering how painful life is. His answer is unbelievably weak! He says that we shouldn't, because we will never know that we did it! Well, so what? If our lives are really as horrible as Schopenhauer says, why shouldn't we put ourselves out of our misery? Von Hartmann's reason why we shouldn't is far more interesting! In his main work, *Philosophy of the Unconscious* (1869), he states that the Unconscious comes into consciousness via us. The Unconscious is suffering terribly via us. The reasons we shouldn't commit suicide is that it would fail to annihilate the Unconscious; we would come back again and continue to suffer. Instead, we should affirm life and devote ourselves to social evolution and Science, and thereby discover a way in which the Unconscious could annihilate itself completely! Then, and only then, would suffering cease.

Good God, what a weird idea. But interesting!

Speaking of optimism and pessimism, I heard a tailor joke in two different versions, one with an optimistic and the other with a pessimistic ending, depending on the word order of the last sentence!

A man brings in some cloth and asks the tailor to make him a pair of pants. The tailor tells him it will be ready the next Tuesday. He comes back on the day, but the pants are not ready yet. "Come back next Tuesday," the tailor says. He comes back the next Tuesday and the pants are still not ready. This goes on and on for six months, and then the day arrives when the pants are ready. "You know," said the man, "that it took our Lord six days to make the entire world, and it takes you six months to make a pair of pants!" Now, in the optimistic ending, the tailor begins "But look at the world," and continues with a bright happy face, "and look at the quality of these pants!" In the pessimistic ending, the tailor begins with a happy smile: "Look at the quality of these pants!" and continues with a very sad face "and look at the world!"

Another tailor joke: A man brought in some material to a tailor and asked him if there was enough to make a suit. The tailor brought the material to the back of the shop, measured it, came back and said: "Sorry, not enough material." The man walked out, took the material to another tailor a block away, and asked him whether there was enough material for a suit. After measuring it, the tailor said: "Yes, there is enough," and told him when to come back for the suit. The man did come back on that day, and there was the suit, beautifully made. The tailor said: "You know, I should really have asked your permission first, but there was so much material left over, that I made a garment for my little boy. Of course, I will pay you for the material, and I hope you will excuse me." The man said: "You really did such a fine job on my suit, that you are perfectly welcome to the extra material."

He then went back to the first tailor and said: "I don't understand! You tell me that the material I brought you was not enough for a suit, and I bring the same material to your competitor down the block, and he not only has enough for a suit for me, but also for a suit for his boy. How come?" The tailor replied: "So can I help it that my boy is bigger than his boy?"

It is interesting how unreliable are the accounts of some people. For example, I once did a magic trick to a girl involving three little paper balls. At one point I appeared to put one of the paper balls into her hand and asked her to close her hand over it, but in reality I had made a switch and actually placed a rolled up dollar bill into her hand. When she later opened her hand, to her great surprise, she found a dollar bill instead of a paper ball. She later reported the trick to a friend and said: "He put a dollar bill into my hand and when I opened it, it wasn't there anymore."

Good Heavens! I don't think *any* magician could do *that* trick!

More seriously, when I taught at the graduate center of C.U.N.Y., on one occasion I was having lunch in the upstairs cafeteria with a group of graduate students, one of whom was an ultra-radical feminist who at one point in our conversation said that the only quality in a person that matters in intelligence. I said: "Oh, come on! Intelligence is certainly a most valuable quality, but surely there are other qualities just as important. What about things like artistic ability, sense of humor, decency of character? I can certainly love someone who is not necessarily intelligent if the person has other valuable qualities." The next day, her roommate met me in the hall and said: "Professor Smullyan, I understand that you like only women who are dumb!"

Good God, what a distortion! I never said anything about women in particular, I spoke about *people*. Also, I never said that I prefer someone who is *not* intelligent.

As Anatole France wisely said, "If a million people say a foolish thing, it is still a foolish thing."

This reminds me of one of my favorite Chinese proverbs: When the wrong person does the right thing, it usually turns out wrong.

I am extremely fond of ancient Chinese writings. As an example, some ancient Chinese philosophers were once discussing whether the human soul was basically good or bad. One of them wisely said: "A human soul is neither good nor bad. It goes through good stages and bad stages, just like a river is sometimes clear and sometimes muddy."

I am also fond of the writings of ancient India. In one of them is discussed the futility of trying to become a sage by imitating a sage. "It is like a jackal imitating a lion. It can imitate it through eons and eons; it will never become a lion."

Of course, Hinduism believes in reincarnation. I like what the poet Heinrich Heine said about reincarnation. He said: "I believe it is a real possibility. I wouldn't be surprised if the soul of Pythagoras right now inhabits the body of the student who just failed his examination because he couldn't prove the Pythagorean theorem."

I also recently came across a passage from the Hindu scriptures in which the philosophers were wondering how the universal came into existence. "The Gods do not know because they came later, but the Supreme One, he knows! Or perhaps he knows not."

Now, that's what I call true modesty. And speaking of modesty, Mark Twain once said: "I was born modest, but it didn't last long!" I also like what Conan Doyle said about modesty through the mouth of Sherlock Holmes. In the story of the Greek Interpreter, Holmes tells Watson that his brother Mycroft is more intelligent than him. Watson tells Holmes: "I think you are being modest!" Holmes replies: "No, no, Watson. I never regarded modesty as one of the virtues. To underrate oneself is just as much a departure from the truth as to overrate oneself."

How wise! The following story is one that not everyone appreciates, but some do: The story is about the world's most modest man. He signed all his letters as "He who is modest." A religion student once said to his instructor: "Now, how can he be modest when the very way he signs his name belies the fact?" The instructor answered: "You don't understand. He is modest. It's just that ever since modesty entered his soul, he no longer regards it as a virtue."

I once thought of the following dialogue about modesty:

> A: For a person of your talents, you are remarkably modest!
> B: I'm not modest!
> A: Ah, I've caught you! By declaiming your modesty, you are trying to show that you are so modest that you won't take credit for anything, not even your modesty. But this is most immodest of you!
> B: It's like I said; I'm not modest.

Someone I know once angrily and hostilely criticized me for being a show-off. I replied: "I certainly am a show-off, thank God! Being a show-off is in itself neither good nor bad, but depends entirely on the quality of what it shows. If the quality is good, then showing off is a benefit all around. If the quality is not good, then showing off can be a bloody bore. Now, I know that I have a talent for entertaining people and it would be very wrong of me not to use it!"

The above incident is an example of a case where someone compliments me intending to do the opposite. Here is another example.

When I was in high school, one of my heroes was Bertrand Russell. I had one English teacher who was highly cultured, very well read, but extremely conventional. On one occasion we had to write essays. I wrote one in which I expressed very radical ideas. She returned me the paper with a comment: "Your English is good, but your thinking is confused. Please see me in my office." Well, the next day I went to her office and we had a long chat about various authors. At one point I asked her what she thought of Bertrand Russell. She angrily said: "He's like you! His thinking is confused!"

Boy, did I feel good! Curiously enough, I was once reading Bertrand Russell to my mother, and at one point she said: "You know, he reminds me of you."

Someone once asked Russell what was really new in the conclusion of a syllogism, since all of its information was already contained in the premises. Russell answered that *logically* there was nothing new, but the conclusion can certainly have psychological novelty, and illustrated this with the following story: At a party, someone once told a risqué joke. Someone else said to him: "Ah, ah! Be careful! Don't you realize the Abbot is here?" at which the Abbot said: "Oh, we men of the cloth are no so naive. We have seen much of life. Why my very first penitent was a murderer." Shortly after, a certain Lord So-and-So arrived and was asked if he knew the Abbot. He replied: "Why of course I know him. Indeed, I was his first penitent."

The one who introduced me to the writings of Bertrand Russell was my cousin, the philosopher Arthur Smullyan, who was a keen arguer. In one philosophy course he took, he kept arguing and arguing with the professor. The professor finally said to him, "Now look, Mr. Smullyan, this is a history of philosophy course, and I want no more argumentation. If you want to ask questions, I will answer them, but please, no more argumentation." To which my cousin replied: "Very well. Then I would like to ask a question: How would you answer the following argument?"

Speaking of questions:

> A: Can I ask you a question?
> B: Yes.
> A: How much is two plus two?
> B: Four.
> A: And how much is two times two?
> B: Four.
> A: How much is two minus two?
> B: Zero.
> A: Now tell me how many questions I asked you.
> B: Three.
> A: Wrong!
> B: Why?
> A: The first question I asked you was "Can I ask you a question?"

Speaking of arguments, a man once said to an elderly man: "You look remarkably healthy for a man your age. What is your secret?" The elderly man replied: "It's because I never argue." The man said: "Oh, I bet you sometimes argue." The elderly man replied: "Maybe you're right."

I thought of the following dual version of the above. One man says to another: "I think it is alright to argue sometimes." The other says: "I disagree!"

Now for some of my favorite limericks. The first two are real tongue twisters when spoken fast. (When I tell these limericks to people, I first tell them very fast and then I repeat them very slowly so they can understand them. I believe this is the way tongue twisters should be told.)

A tooter who tooted the flute
Tried to tutor two tooters to toot.
  Said the two to the tutor,
  Is it harder to toot,
Or to tutor two tooters to toot?

A canner exceedingly canny
One morning remarked to his granny,
  A canner can can
  Anything that he can,
But a canner can't can, can he?

There was a man from Peru
Who dreamed he was eating his shoe.
  He woke in the night
  With a horrible fright
And found it was perfectly true!

Speaking of tongue twisters, a man who once read one, told a friend that he didn't like it. When asked why, he thought for a while and finally said: "It's hard to say."

Now for some more jokes: A man told his friend that his grandfather knew the exact date and time that he would die. When the friend asked him how he knew, he said: "The judge told him."

A woman went to a fortune teller who looked into her crystal ball and said: "Within one year, your husband is going to die a violent death!" The woman appeared quite frightened and asked: "Will I be found out?"

JUDGE: And therefore the court sentences you to be hanged by the neck until you are dead.
DEFENDANT: That's all I need!

Jack and his friend went cross-country skiing. They got lost in a blinding snowstorm. They came to a farm house and knocked on the door. It was answered by a beautiful woman. They asked whether they could stay for the night. The lady told them that she was recently a widow and it wouldn't look right to her neighbors if they stayed in her house. They then asked her if they could sleep in the barn. She said that would be alright. And so they slept in the barn and left early the next morning.

A few months later, Jack received a letter from the lady's attorney. Puzzled, he went to his friend and said: "Now, tell me honestly. Did you leave the barn that night and go into the house and visit the lady?" Embarrassed, he said: "Well, yes I did." "And when she asked you your name, did you give her my name instead of yours?" He laughed and said: "Why yes I did, why do you ask?" "She just died and left me everything."

JUDGE: (to defendant) You mean to say that you killed this helpless old woman for a paltry three dollars?
DEFENDANT: Well, your Honor, three dollars here, three dollars there, it mounts up!

A man stopped another man in the street and asked how to get to a certain place. The other replied: "You go uptown two blocks and then turn left, no, no, you first go downtown and turn left and, oh no, you can't get there from here."

An American in Berlin was trying to find the opera house and asked a Berliner how to get there. The Berliner gave him a long, elaborate set of directions. The man said: "Thank you." The Berliner said: "Vat do you mean, *Thank you*? Repeat!"

Here are some puzzles (answers given later).

1. A man was driving along a highway. His headlights were broken, there were no street lights on and there was no moon out. There was a pedestrian crossing the street about a hundred and fifty yards in front of him. The driver knew that the pedestrian was there and stopped his car in time to avoid hitting him. How did he know that the pedestrian was there?
2. A man held in his hand two American coins that added up to 30 cents, yet one of them was not a nickel. How do you explain that?

3. What is it that's greater than God, the dead eat it, and if the living eat it, they die?

Answers:

1. It was daytime.
2. He held a quarter and a nickel. One of them—namely the quarter—was not a nickel.
3. What do the dead eat? Obviously, nothing. Hence "nothing" is the answer. Nothing is greater than God, the dead eat nothing, and if the living eat nothing, they die.

Some time ago I came across a cartoon showing a window cleaner outside a window high up in an apartment building, strapped in by a seat belt, saying to himself: "I must remember to get me a new seat belt; this one is about shot."

I recall an old-time joke about two friends walking in the street and suddenly accosted by a holdup man. One of the friends said to the other: "Here is the hundred dollars I owe you."

In one old-time comedian act, a comedian said to the other: "You're a good friend of mine, aren't you?" "Sure!" "If you had two million dollars, you'd give me one million, wouldn't you?" "Sure!" "And if you had two shirts, you would give me one, wouldn't you?" "No." The other said, "Why not?" He replied: "I have two shirts."

Another comedian said, "I'm very much like my old man. He wouldn't think anything of getting up at four o'clock on a cold morning and going out for a walk. I don't think anything of it either."

Another old-time comedian in his act said: "And it's pretty tough these days when you go into a restaurant and have to pay two dollars for a steak! If you pay one dollar, it's even tougher!"

In another comedian act, the comedian said to his friend: "I defended you the other day. Someone told me that you weren't fit to live with a pig, and I said you were."

Speaking of pigs, do you know what kind of mathematics pigs study? Why, pigs study pigonometry, which deals with swines and coswines.

Another old-timer: A man went over to another man and said: "Why Joe, how you have changed! I see you've dyed your hair, you have gained weight and you even look taller!" The other replied: "My name is not Joe!" The man said: "Oh, so you even changed your name!"

The following is true: A six-year-old boy was at a banquet. At one point the hostess said to him: "Would you like some more chicken?" The boy abruptly said: "No." His mother said to him: "No *what* dear?" The boy replied: "No chicken."

There is the joke told about a man who told his friend that he was invited to a banquet and that he knew that many girls would be there, but he didn't know how to talk to girls. "If I should sit next to a girl," he asked, "what do I say to her?" The friend told him, "Well, first talk about something commonplace like food. Then show the social side of yourself and talk about family. Then show the intellectual side of yourself and talk about philosophy." "Okay," he said, "food, family, philosophy. Food, family, philosophy. I'll remember that."

Well, at the banquet he was indeed seated next to a girl. After some silence he asked her: "Do you like cabbage?" She replied, "No." After some more silence, he asked: "Do you have a brother?" She replied, "No." After yet more silence he asked her: "If you had a brother, would he like cabbage?"

Now let me ask you a question: Which is better, eternal happiness or a ham sandwich? Most people of course say eternal happiness, but this cannot be true, as the following syllogism proves:

> Nothing is better than eternal happiness.
> A ham sandwich is better than nothing.
> Therefore, a ham sandwich is better than eternal happiness.

Speaking of ham sandwiches, I was recently in a restaurant and met a lovely and highly intelligent waitress whom I will call "Betty." She was a college senior, and as you will see, she had the perfect makings of a very good lawyer!

When I first saw her, I asked her: "How come you are looking so beautiful?" She replied: "Because you are here." After having flirted with her during dinner, just before I left I asked her: "Have you ever met any man as dangerous as me before?" She replied: "Yes, but not as suave."

Two days later I came to the same restaurant for lunch with a friend. I said to Betty: "Today I am even more dangerous than last time!" She replied: "That's unimaginable!" Before I left, I looked at her wistfully, sighed, and said: "Ah, if only I were sixty years younger!"

At the end of the summer she told me that she now had to go back to school and wouldn't be back here for a couple of months. I said: "Then I'll have to live with Betty deprivation?" She replied: "I think you'll survive." I wasn't sure I heard what she said, and asked: "Did you say that you will survive or that I will survive?" She replied: "I said that you will survive," and turned her head and added: "I don't know if I'll survive!"

A couple of months later, when she came back, she sat opposite me in the restaurant, and with an impish expression said: "When are you and I going to elope?" Of course she was kidding, but as Shakespeare said, "Many a truth is said in jest."

Now, doesn't she sound as if she would be an excellent lawyer? I suggested to her that she go to law school instead of business school, which she had planned. She took my suggestion and is now a law student, and, I understand, is doing extremely well!

I am reminded of the joke in which a fairy came to the husband of a middle-aged couple and told him that she would grant him one wish. He said: "I want a wife thirty years younger than me." She said: "Alright," and waved her magic wand, and he suddenly became thirty years older.

Now for some more logic. I recently came across a logic puzzle that I think is highly instructive and psychologically revealing. I will also show you some interesting variations of the puzzle.

In the original problem, Arthur is looking at Betty who is looking at Charles. Arthur is married and Charles is not. Has enough information been given to determine whether a married one is looking at an unmarried one?

I sent this problem to the logician Professor Robert Cowen (a former student of mine) who thought of a nice generalization: You have a line of people. The first one in line is married and the last one is not. Prove that at least one married one is directly behind an unmarried one.

Then I thought of the following variation of his variation. Again there is a line of people. At least one of them is married and at least one is not. Prove that either some married one is directly behind an unmarried one, or some unmarried one is directly behind a married one.

Solutions: For the first problem, about 80 percent of people say that there is not enough information, because there is no way of knowing whether Betty is married or not. It is indeed true that it cannot be determined whether or not

Betty is married, nevertheless there is enough given information to imply that a married one is looking at an unmarried one. Either Betty is married or she is not. If she is, then married Betty is looking at unmarried Charles. If she is not married, then married Arthur is looking at unmarried Betty. In either case, a married one is looking at an unmarried one.

The interesting thing about the above argument is that it is what mathematicians call *non-constructive*, in that it does not tell you *which one* is the married one looking at an unmarried one; it merely shows that it is either Arthur or Betty, but there is no way to know which.

The author of this problem called it an example of *disjunctive reasoning* and stated that most people carry out disjunctive reasoning when they are explicitly told that it is necessary, but most do not automatically do so, and the tendency to do so is only weakly correlated with intelligence.

I found those remarks quite interesting, since I know two extremely intelligent people who did not get the solution and one not so intelligent person who did.

For Cowen's generalization, the last person in line who is married must be directly in back of an unmarried one.

For my variant, the last one in line whose marital status is different from the first one in line—that person is directly in front of one of different marital status.

Here now is a group of logic puzzles that I call Gödelian puzzles, since they are related to a very famous result in mathematical logic known as Gödel's theorem, which I will later tell you something about.

a) Suppose I put a penny and a quarter on the table and ask you to make a statement. If the statement is true, then I promise to give you one of the coins, not saying which one. But if the statement is false, then I give you neither coin. What statement could you make such that I had no option other than to give you the quarter (assuming I kept my word)?

b) There is another statement you could make such that I would have to give you both coins. What statement would work?

c) There is yet another statement that would make it impossible for me to keep my word. What statement would that be?

d) For many years I have presented this problem to a logic class, until one day I realized to my horror that the student could make a statement such that the only way I could keep my word is by paying him a million dollars! What statement would work?

Answers:

a) A statement that works is: "You will not give me the penny." If the statement were false, then contrary to what it says, I would give you the penny, but I can't give you the penny for a false statement (according to the agreement). Therefore the statement can't be false; it must be true. Since it is true, then what it says is really the case, which means that I don't give you the penny. Yet I must give you one of the two coins for a true statement, hence I just give you the quarter.

b) A statement that would work is: "Either you will give me neither coin or both coins." The only way the statement could be false is that I give you one coin but not the other, but I can't give you a coin for a false statement, and therefore the statement can't be false; it must be true. Thus it is true that I give you neither or both, but I can't give you neither, since the statement is true, hence I must give you both.

c) A statement that obviously works is "You will give me neither coin." If I give you a coin, then it is false that I give you neither coin, hence I gave you a coin for a false statement and I have broken my word. On the other hand, if I don't give you a coin, then it is true that I give you neither coin, hence I failed to give you a coin for a true statement, and so again I have not kept my word.

d) All the student needed to say is "You will give me neither of the two coins nor a million dollars." If the statement were true I would have to give him one of the two coins, which would falsify the statement and we would have a contradiction. Therefore the statement can't be true; it must be false. Since it is false that I give him neither one of the coins nor a million dollars, I must give him either one of the two coins or a million dollars, but I can't give him either coin for a false statement, hence I have no option other than to give him a million dollars!

Here is another group of puzzles that are even more closely related to Gödel's theorem.

1. A certain logician named Godwin was one hundred percent accurate in his proofs, in that everything he proved was really true. He once visited a very strange land in which every inhabitant was either truthful and always told the truth, or always lied. Godwin met an inhabitant named Theodore who made a statement from which it follows that Theodore must be truthful, but the logician Godwin could not possibly prove that he is. What statement would work?
2. In this land, certain truthful ones had been proven truthful, and they were known as *certified* truth tellers. Well, a certain inhabitant made a statement from

which it logically follows that he must be truthful, but not certified. What statement would work?

3. This is a side issue, but in a more complicated situation, the logician came across two inhabitants named Arthur and Bernard who each made a statement from which it logically follows that one of the two must be an uncertified truth teller, but there is no way to tell which one it is. From neither statement alone could that be determined. What two statements would work?
4. A certain system, which we will call system S, proves various English sentences. The system is wholly accurate in that everything provable in the system is really true. There is a sentence which must be true and yet it cannot be proved in system S. What sentence could that be?

Answers:

1. A solution is that Theodore said: "You can never prove that I am truthful." If he were untruthful, then, contrary to his false statement that the logician could not prove that he is truthful, the logician *could* prove that he is truthful, which the logician cannot do, since he proves only true facts. Therefore the sentence can't be false; it must be true and so Theodore is truthful. Since his statement that the logician cannot prove that he is truthful is true, then the logician really cannot prove that Theodore is truthful. Therefore Theodore is truthful, but Godwin can never prove that he is.
2. One solution is that he said: "I am not certified." He couldn't be lying, because if he were, then contrary to what he said, he would be certified, but no liars

are certified. Therefore he told the truth, hence he is not certified as he truthfully said. Thus he is an uncertified truth teller.

3. Here is one solution: They said the following:

    ARTHUR: Bernard is certified.
    BERNARD: Arthur is not certified.

   To prove that one of them is an uncertified truth teller, we must divide the proof into two cases—either Arthur is truthful or he isn't.
   Suppose Arthur is truthful. Then Bernard is certified as Arthur said. Hence Bernard is of course truthful, hence Arthur is uncertified as Bernard said. And so in this case Arthur is an uncertified truth teller.
   Now consider the case that Arthur lied. Then contrary to what he said, Bernard is not certified. Yet Bernard told the truth that Arthur is not certified, since no liar can be certified. And so in this case, Bernard is an uncertified truth teller.
   In summary, if Arthur is truthful then he is an uncertified truth teller, and if he lied, then Bernard is an uncertified truth teller. There is no way to tell which one it is.
   Remarks: I am particularly fond of this type of puzzle in which it can be shown that one of two possibilities must hold, but there is no way to tell which. Another one of mine of this type is in my book *What is the Name of This Book?* in which I gave a proof that either Tweedledee or Tweedledum exists, but there is no way to tell which. It is puzzles of this type which led the logician Melvin Fitting to once introduce me at a math lecture by saying: "I now introduce Professor Smullyan, who will prove either he doesn't exist, or you don't exist, but you won't know which."

4. A sentence that obviously works is: "This sentence is not provable in system S." If false, then contrary to what it says, it would be provable in system S, which cannot be since only true sentences are provable in that system. Hence the sentence is true, and as it says, it is not provable in system S.

I told you that the puzzles we just considered were related to Gödel's theorem, which I will now tell you something about.

Until the year 1931 it was taken for granted that the most powerful known axiom systems at the time were powerful enough to decide all mathematical questions—any mathematical statement could supposedly either be proved or disproved in the system. In 1931, Kurt Gödel startled the mathematical world by showing that this was not the case. He showed that for these systems, as well as a wide variety of related systems, there must always be a sentence which, though true, could not be proved in the system. What Gödel did was to assign to each sentence of the system a number called the *Gödel number* of the sentence, and then he very ingeniously constructed a sentence G that asserted that a certain number n was the Gödel number of a sentence which was not provable in the system, but the number n was the Gödel number of the very sentence G! Thus G asserted that its own Gödel number was the Gödel number of an unprovable sentence, which is tantamount to asserting its own non-provability. This means G is either true, but not provable (as it says), or false, but, contrary to what it says, *is* provable. Thus G is either true and unprovable, or false and provable. Under the reasonable assumption that the system is correct in that no false sentences are provable, the second alternative doesn't hold, and so Gödel's sentence G is true but not provable in the system.

Incidentally, the way my first problem about the penny and the quarter was suggested to me by Gödel's theorem is this: I thought of the penny as standing for provability and the quarter as standing for truth, and so giving the quarter and not the penny corresponds to the sentence which is true but not provable.

Speaking of Gödel numbering, there is the well-known story about numbering of jokes, and this story has two different endings, one of which is well known, and the other is much less known but I believe much funnier! The joke is about the president of a joke makers convention who invites a friend to a banquet of the joke makers. The friend came and was quite puzzled at what went on: several times during the banquet a member would arise and call out a number and everyone would laugh. When the friend asked the host the meaning of all this, the host explained: "We joke makers don't want to take the time to tell the whole joke. Instead we have our jokes numbered, and so one of us calls out the number of a joke, and this gets us to recall the joke and we then laugh." Now, in the well-known variant, the friend asks the host whether he can try it. The host agrees. The friend then stands up and calls out a number. No one laughs. The friend asks the host why no one laughed. The host says: "Some people can tell a joke and some people can't."

That is the well-known ending. The other, which I like far better, is that one member got up and called out a very high number. Everybody laughed and one person kept laughing long after everyone else stopped. The friend asked the host: "Why is he still laughing?" The host replied: "Oh, he hasn't heard that one before."

Okay, enough jokes! Let me conclude by telling you of a particular interest of mine, Eastern culture—particularly ancient Chinese philosophy, poetry and paintings, which were then regarded as essentially one and the same thing. One thing I love about Chinese writings on the theory of art is that, unlike Western aesthetic criticism, the writings have the same kind of beauty as the paintings they describe. As to their poetry, I particularly love that known as the poetry of the recluse.

One of the earliest is that of T'ao Chi'en (3rd century):

> I built my house near other's dwelling
> No clamor of carriage nor horse
> I pluck chrysanthemums from the eastern hedgerow
> And gaze towards too distant mountains
> The mountain air is fine in the evening cool
> A bird flying together homeward
> Within all these things, there is a hidden truth
> But when I try to tell it, I get lost in no words.

The above is a composite translation from several sources, as is the following one. This next is, I believe, quite profound. It is about an old tree.

> In the forest lies a tree
> As old as the forest itself
> Its years of life cannot be reckoned
> Fools laugh at its shoddy exterior
> Not knowing that stripped of its bark,
> Lies the core of truth!

Some other composite translations I made of Chinese poems can be found in my books *Rambles Through My Li-*

*brary** and *The Tao is Silent*†, which contains ideas inspired by Taoism and Zen Buddhism. The original edition contains a little poem of mine (which, curiously enough, was left out in later editions) titled "The Tao is a Silent Flower," which is a poetic expression of existence arising out of nonexistence:

> The Tao is a silent flower
> Which blooms through the night.
> But the night through which it blooms is the
>   flower itself.
> No Tao, no flower, no blooms, no night
> And for this reason it blooms.

Oh, one other thing. I must tell you of a certain great Sage in the East who was reputed to be the wisest man in the world. A philosopher heard about him and was anxious to meet him. It took him fifteen years to find him, but when he finally did, he asked him: "What is the best question that can be asked, and what is the best answer that can be given?" The great Sage replied: "The best question that can be asked is the question you have asked, and the best answer that can be given is the answer I am now giving."

---

*Praxis International Inc, 2009.

†Harper and Row, 1977

About the Author

Raymond Smullyan, born 1919 in Far Rockaway, New York, is a mathematician, concert pianist, magician, and author of numerous books of logic puzzles, chess puzzles, mathematics, philosophy and memoir. His startlingly brilliant and original puzzle books, of a genre largely invented by him, are both delightfully approachable and devilishly challenging, leading the reader by degrees from elementary puzzles to deep results in mathematical logic. His soulful piano playing of works by Bach, Schubert and Scarlatti, among others, are on a high artistic and spiritual level. As a person as well as a creator, he is *sui generis*; part Taoist sage, part *bon vivant*, he is a natural born entertainer who finds an audience to charm wherever he goes, performing magic tricks, telling jokes, spreading wisdom and good cheer, and inspiring more often than not the comment "You've made my day!" He lives in the Catskill Mountains.